PREPARATION OF NUCLEAR TARGETS FOR PARTICLE ACCELERATORS

PREPARATION OF NUCLEAR TARGETS FOR PARTICLE ACCELERATORS

Edited by

Jozef Jaklovsky

New England Nuclear Corporation
Boston, Massachusetts

PLENUM PRESS • NEW YORK AND LONDON

Library of Congress Cataloging in Publication Data

World Conference of the International Nuclear Target Development Society (1979: Boston)

Preparation of nuclear targets for particle accelerators.

"Proceedings of the World Conference of the International Nuclear Target Development Society, held October 1-3, 1979, in Boston, Massachusetts."

Bibliography: p.

Includes indexes.

1. Targets (Nuclear physics)—Congresses. I. Jaklovsky, Jozef. II. International Nuclear Target Development Society. III. Title.

TK9340.W67 1979 539.7'3 81-7293

ISBN 0-306-40731-0 AACR2

Proceedings of the World Conference of the International
Nuclear Target Development Society, held October 1 – 3, 1979,
in Boston, Massachusetts

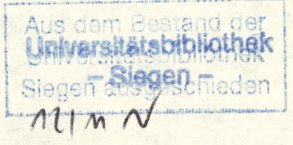

Printed in the United States of America

PREFACE

This book contains the Proceedings of the World Conference of the International Nuclear Target Development Society held in Boston, October 1-3, 1979. Approximately seventy participants and registrants from the following countries attended this Conference: Belgium, Canada, Denmark, United Kingdom, German Federal Republic, Israel, Japan, The Netherlands, Saudi Arabia, Switzerland, and the United States.

The Conference was sponsored by the New England Nuclear Corporation. Conference Chairman in residence was Jozef Jaklovsky who was assisted by the invited Co-Chairman D. Riel of the State University of New York at Stony Brook.

This publication incorporates the proceedings of an international conference concentrating on the general field of target preparation for use with particle accelerators. The publication includes additional contributions on target preparations by honorary members of this Society. For the first time in this field, an extensive bibliography collection (1936-June 1980) is included. Bibliography, subject index, author index, country index, and patent index have been organized by the editor. The bibliography includes thin and thick particle accelerator targets and also preparation and use of targets in particle accelerators.

The Chairman of the 1979 World Conference would like to express his gratitude to the International Nuclear Target Development Society for the decision to hold its' 8th Annual meeting at the New England Nuclear facility in Boston, Massachusetts. Thanks are also given to the session Chairmen (G. E. Thomas, E. K. Kobisk, J. Van Audenhove, P. Maier-Komor, and J. L. Gallant), the speakers for their excellent contributed papers, and to all participants who contributed invaluable ideas to the discussions. A special note of gratitude is given to my wife, Jolana, and to Bill Evans and Stephen Talutis for helping me to organize this bibliography.

The contributions of the following industrial companies enabled us to make this Conference highly successful:

New England Nuclear
Oak Ridge National Laboratory
Edward Lead Company
Central Research Labs, Inc.
Excel Metallurgical
Fisher Scientific
Packard Instrument Company, Inc.
Microwave Techniques, Inc.
Fairview Machine Co., Inc.
Maurer and Sforza
Carl Zeiss, Inc.

The contributions of these companies are greatly appreciated.

Thanks are given to Plenum Publishing Corporation for publishing these Proceedings.

Billerica, Massachusetts Jozef Jaklovsky

CONTENTS

CARBON STRIPPER FOILS

SPECIAL TARGET PREPARATION METHODS

GENERAL TARGET PREPARATION METHODS

OTHER TOPICS RELATED TO TARGET

LIFETIME MEASUREMENTS ON CARBON STRIPPER FOILS

R. L. Auble and D. M. Galbraith

Oak Ridge National Laboratory*
Oak Ridge, TN 37830

One of the most exciting recent developments in accelerator technology is the new breed of electrostatic accelerators. These machines have terminal potentials of 25 million volts and higher and are being built to accelerate intense beams of heavy ions. One such accelerator is now nearing completion at ORNL. A crucial component in these--as in any electrostatic accelerator--is a tiny bit of carbon in the form of a very thin foil. These foils, called stripper foils, are used to strip electrons from atoms so that they will be accelerated to high energy by the electrostatic potential. While carbon foils are not the only method of stripping electrons, they have a number of advantages over the alternative gas strippers. In particular, they produce higher charge states which are essential for providing the energetic ions required for heavy ion physics experiments. Because of the importance of stripper foils to the most efficient operation of the new ORNL accelerator, we are quite interested in their properties and in improving their performance.

Carbon foils have been in use for many years in smaller accelerators. One property of these foils that has been well established is that their performance deteriorates under bombardment by energetic ions. One of the reasons for the deterioration is that the foils shrink parallel to the foil in the irradiated area. This causes an effective thickening of the foil resulting in increased multiple scattering and loss in extracted beam

*Research sponsored by the Division of Basic Energy Sciences, U. S. Department of Energy under contract W-7405-eng-26 with the Union Carbide Corporation.

intensity. The shrinkage also results in stresses in the surrounding unirradiated foil which can lead to mechanical failure of the foil. Since the rate of shrinkage increases for heavier projectiles and higher beam currents, stripper foil failure can represent an important limitation on the efficient operation of the new and costly accelerators.

Until recently, stripper foils had traditionally been made by vapor deposition techniques. In this method, the carbon is vaporized by a carbon arc or electron gun heating and collecting the carbon on coated glass substrates which are normally near room temperature. Recently, Takeuchi and coworkers in Japan reported dramatic improvements in vapor deposited foils by using carefully prepared heated substrates. Similar improvements have also been

Figure 1. The glow discharge chamber used at ORNL.

reported recently by researchers at Daresbury in the U. K. using a quite different technique. In the Daresbury experiments, foils were produced by cracking hydrocarbon gases in a glow discharge with carbon being deposited on the substrate which served also as the cathode. While this technique had been used earlier to produce adherent films, this was the first successful production and testing of self-supporting foils. These successes led to the current ORNL efforts on carbon foil production and testing.

Initial screening experiments were begun in March, using foils supplied by J. L. Gallant from Chalk River Laboratories in Canada. Gallant had successfully produced and tested foils using both the hot substrate and glow discharge techniques and kindly provided specimens to be tested on the ORNL EN tandem accelerator facility. These tests were performed in March using a 10 MeV chlorine beam. In these preliminary tests, we found that both types of foils had lifetimes which were an order of magnitude or more longer than our conventional foils. Since both types appeared to have comparable lifetimes, the decision was made to develop a production facility at ORNL based on the glow discharge technique. The choice of technique was based on its inherent simplicity and the likelihood of highly reproducible results.

The glow discharge chamber used at ORNL consists of a modified dessicator jar and is shown in Figure 1. Here one can see the inlet port for the gas mixture, the vacuum pumping port, and the electrodes. It was found that a separate anode was not required and thus the aluminum top serves as both the anode and the vacuum seal. The cathode, on which the substrate is placed during the glow discharge, is a copper plate supported by a teflon insulator. The cathode is connected to the negative high voltage supply through a 5000-ohm ballast resistor which suppressed arc discharges. Since the volume of the chamber is quite small, a continuous flow of gas is maintained during the discharge. Following evacuation of the system with a diffusion pump, a mechanical pump is opened to the system and the gas flow adjusted to give a pressure of 0.1 torr in the chamber, the pressure being that on a capacitance gauge connected to the vacuum line. The glow discharge is then initiated by applying bias to the cathode for the desired length of time. During the glow, the pressure rises to about 0.13 torr due to release of hydrogen from the ethylene.

The substrates used for the glow discharge apparatus are 3" x 3" squares of 10-mil stainless steel which is chrome-plated and polished on one side. The substrates are washed in alcohol and dried and then placed in a vacuum evaporator where they are coated with sodium chloride as a release agent. They are then immediately transferred to the cathode of the glow discharge apparatus and coated with carbon using glow times ranging from 15-80 seconds, and voltages of 2 to 2.8 kV.

Figure 2. Die for the preparation of slackened foils with mounted foils before and after the reduction process.

For the first foils made, we attempted to mount them by the usual floating on water and subsequent pickup method. We found quickly that these foils are quite fragile and the success rate in mounting them was very poor. Consequently, we adopted the collodion coating process as used at Chalk River to provide support for the carbon foils during mounting and drying. No attempt was made to remove the collodion prior to irradiation testing although this can be done easily using a heat lamp or flash gun. In addition to foils mounted on conventional holders, several were mounted on aluminum rings and slackened using the technique reported by the Daresbury group. Figure 2 shows the die used and foils before and after the reduction process.

Irradiation testing of ORNL foils was done in August using, again, 10 MeV chlorine ions. Foils from nine separate glow discharges, as well as conventional vapor deposited foils were used in these tests. The main criteria for the foils was that they be as thin as practical, namely less than 10 $\mu g/cm^2$, since this is the thickness range of interest for stripper foil applications. The choice of 10 MeV chlorine ions as projectiles is a reasonable compromise between the need for high damage rates and hence reasonable irradiation times and the desire for particle energies which are representative of those which will be experienced by the first stripper foils in the new accelerator. Irradiations were carried out in a high vacuum chamber at pressures in the range of 5×10^{-8} torr, using a cryopump and liquid nitrogen cooled trapping to reduce the deposition of contaminants on the foils due to thermal cracking of residual hydrocarbons. The beam

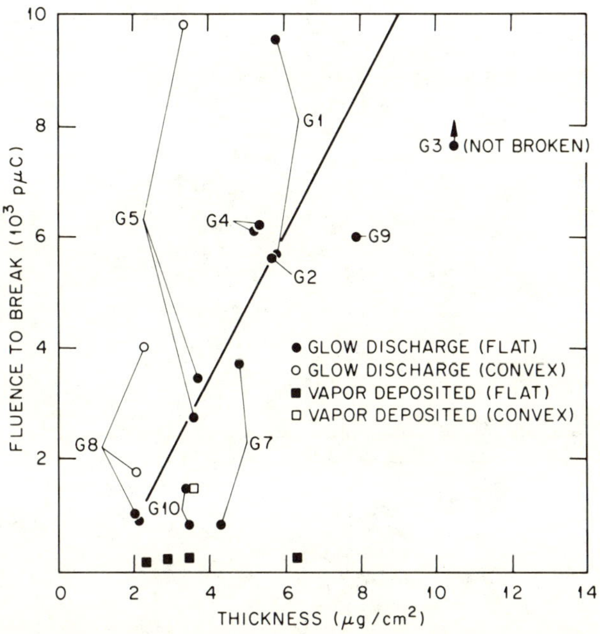

Figure 3. Lifetimes of glow discharge foils of different thicknesses.

focus was adjusted to illuminate an area of about 2-3 mm^2 using a quartz phosphor immediately in front of the foil position. The beam current was monitored by an electron suppressed Faraday cup, and the relative foil thicknesses were determined by detecting recoiling carbon atoms with a surface barrier detector mounted at an angle of 50° to the beam line.

The criterion used to define foil failure in these experiments was that the foil suffer a mechanical failure, that is, that a hole be formed somewhere on the foil. The time of failure was determined by visual inspection of the foil at frequent intervals, although with the thicker glow discharge foil, which ran for 8-10 hours, there was a strong tendency to reduce the number of trips between 2AM and 8AM!

The results of our measurements are summarized in Figure 3. It is apparent that there is a strong thickness dependence present for the glow discharge foils which is not evident or is much suppressed for the vapor deposited foils. Part of this effect could be thermally induced since the beam energy loss causes the thicker foils to be operated at a higher temperature. The calculated temperatures ranged from 300°C for the foils in the 2 μg/cm^2 region to 600°C for the 10 μg/cm^2 foil, and indeed the

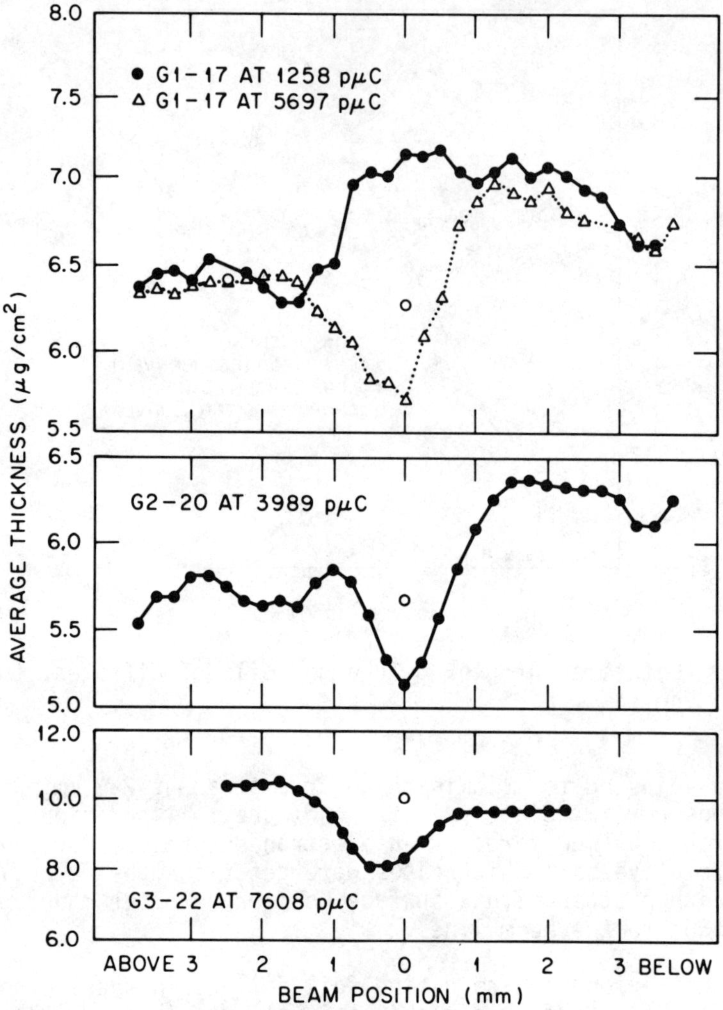

Figure 4. Thickness of three different glow discharge foils, after irradiation to stated fluences, as a function of beam position.

latter foil was found to be faintly luminous at the center of the beam spot. Another factor which may enter into the thickness dependence is the thinning which takes place in the latter stages of the irradiation. Figure 4 shows the variation in foil thickness observed by scanning the beam vertically across the foil. The open circles show the initial thickness measurement and it is clear that significant changes are occurring. In particular, foil 17 was scanned at two different fluences. The first scan was taken at relatively low fluence and shows a significant increase

Figure 5. Carbon foil, which failed at the center of the beam spot.

Figure 6. Carbon foil rupture by tearing in the peripheral region.

in thickness in the irradiated zone. A subsequent scan at high fluence shows the thinning which subsequently occurred and the thickness is in fact less than its initial value by some 9%. The thinning may weaken the foil to the point where it fails at the center of the beam spot. Such a failure is shown in Figure 5. It was also found that several foils, all having relatively short lifetimes, failed by tearing in the peripheral region and such a foil is shown in Figure 6.

Subsequent to the testing, after the relative foil thicknesses were determined, we tried to correlate the thicknesses with the parameters used in the glow discharge. Our initial attempt at

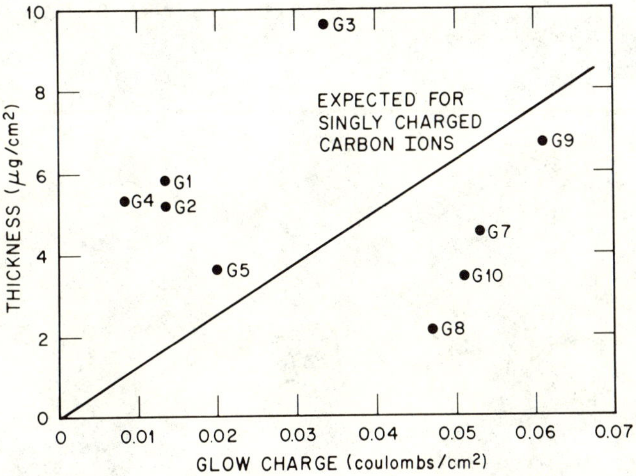

Figure 7. Carbon foil thickness as a function of the charge collected at the substrate.

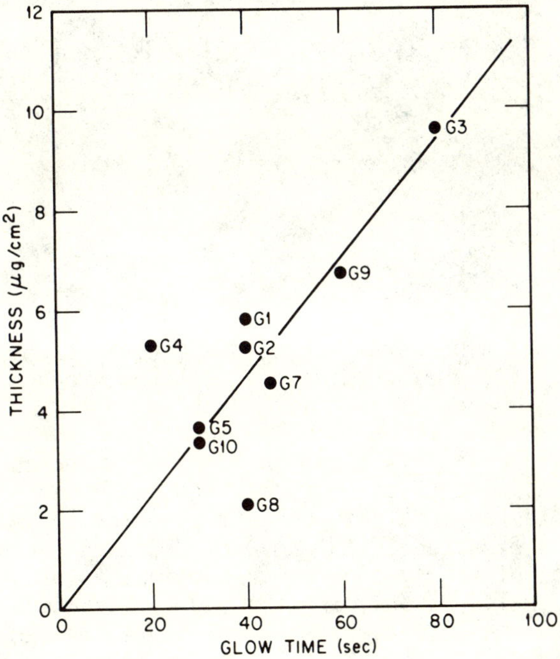

Figure 8. Carbon foil thickness as a function of glow discharge time.

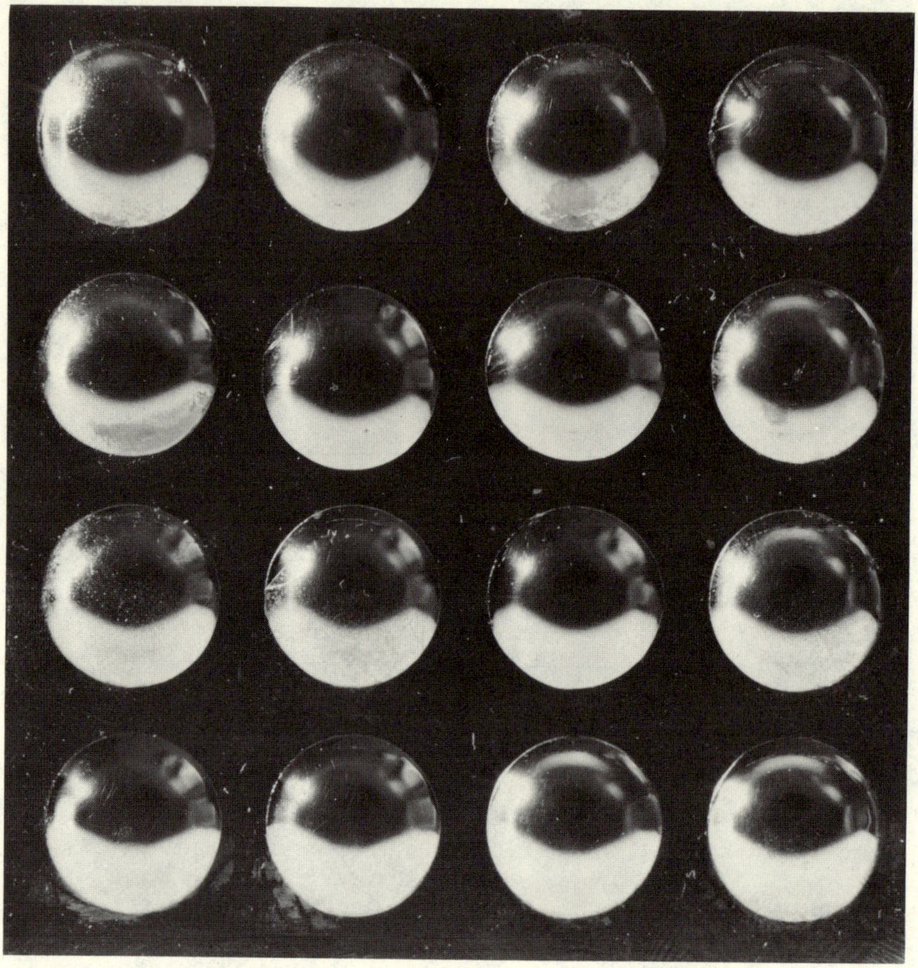

Figure 9. Stainless steel substrate used to prepare slackened foils during the glow discharge process.

correlating these data is shown in Figure 7 which shows the thickness plotted as a function of the charge collected in the glow process, and there appears to be little correlation with behavior which would be expected if the collected charge corresponded to the collection of singly-charged carbon ions. In our second attempt, we considered only the time during which the glow was maintained and the result is shown in Figure 8. Except for some discrepancies at the low end, there seems to be a reasonable correlation between the foil thickness and the glow time. We are planning to study this correlation using well controlled pressure, time, and voltage parameters.

Figure 10. Pressing tool to prepare substrates for slackened foils.

Recently, we discovered that slack foils could be made using a procedure which eliminates the ring reduction process and the problems involved in mounting the rings in the foil changer. In this procedure, the substrate itself is deformed as shown in Figure 9. By using the collodion backing it was found that foils deposited on such a substrate were sufficiently strong that they could be floated off and picked up without destroying their slackened character. Preliminary tests indicate that these foils perform comparably to those made using the ring reduction method.

While the detailed differences in the structure of the foils made by vapor deposition and by the glow discharge technique are not understood at present, it is clear that these new foils represent a timely and valuable contribution to the field of heavy ion electrostatic accelerators.

AN ATTEMPT TO IMPROVE THE LIFETIME OF CARBON FOILS

D. Balzer

Laboratorium fur Kernphysik
Eidg. Technische Hochschule
8093 Zurich, Switzerland

ABSTRACT

The lifetime of carbon foils has been improved by a factor of five by low energy argon ion bombardment. The investigations indicate that this is mainly due to the implanted Ar-atoms and most probably not caused by structural changes in the foil.

INTRODUCTION

The structure of carbon foils is obviously related to the foil lifetime with respect to ion irradiation. Standard techniques of structure analysis, for eg., the X-ray diffraction method, are difficult to apply because of the low atomic number of carbon and the practically minute mass of the foil. Also, a few attempts have been made by electron diffraction studies albeit with inconclusive results. Alterations in the structure are generally inferred from the observation of secondary effects like a change in the electrical conductivity, optical properties, etc. It has, for example, consistently been observed that when a carbon foil is irradiated by an ion beam, initially a certain amount of shrinkage takes place and which ultimately leads to rupture of the foil. This behavior can be attributed to a displacement of the constituent atoms of the material of the foil leading to a disturbance in its structure. In practice, it is almost impossible to determine the structural alterations caused by irradiation. We can however only assume that a certain structure of the carbon - in the form of foils - exists which is more resistant with respect to the effects caused by displacements induced by ion bombardment.

Experiment and Results

The nuclear stopping power[1] is mainly responsible for displacements induced in the irradiated material and has essentially a larger value at low ion energies. Low energy ion beams can therefore appropriately be applied to influence the structure of materials provided the range of the ions is larger than the thickness of the sample under investigation. Furthermore, since the lifetime of foils is inversely proportional to the nuclear collision rate[2,3], such ions are also suitable for measuring foil lifetimes within a much shorter period of time than is conventionally needed by the use of higher energy beams.

Our present efforts have therefore been concentrated in investigating the influence of low energy ions bombarding carbon foils. For the experiments under investigation, the evaporation apparatus was equipped with an ion gun capable of delivering various ion beams in the energy range of 10-30 keV and currents of $\sim$ 20 μA to $\sim$ 2 mA; the rf-source employed produces singly charged ions of all non-corrosive gaseous elements.

In a first set of experiments standard foils* were exposed to beams of helium and argon, both at energies of $\sim$ 10 keV. In the case of helium having the value for the nuclear stopping power S_n calculated according to ref.[1], S_n = 0.069 MeV.mg^{-1}.cm^2, and at a current density of $\sim$ 100 μA/cm^2 the foils ruptured, after an initial slight shrinkage, within an average lifetime of $\sim$ 3 min.

A completely different behavior was exhibited by foils bombarded with Ar-ions having S_n = 3.96 MeV.mg^{-1}.cm^2 and at a current density of $\sim$ 20 μA/cm^2. Due to the larger value of the nuclear stopping power one would expect the foils to have ruptured almost immediately. However, a contradictory observation was made under these conditions. During the first couple of minutes the foils lost their inherent slackness and appeared to be under tension. About 5 min. after the onset of the strain the foils lost that tension, were slack once again although somewhat differently as compared to how they were initially. Ion bombardment could be continued for nearly one hour without rupture to the foil; nevertheless, after this length of time the foil thickness was evenly reduced to zero by sputtering taking place.

*The standard technique[4] employed for the carbon foil fabrication at our lab consists of slow evaporation of the carbon heated by electron bombardment and deposited onto a glass substrate previously covered by an evaporated layer of Cu which serves as the release agent. The foils used for the present investigations were $\sim$ 10-15 μg/cm^2 thick.

The observed tension is similar to the general shrinkage of foils mentioned earlier and can be interpreted as being due to displacements induced under the influence of the radiation. It is more difficult to explain the subsequent release of the tension. This can only be interpreted as being related to the short range of the bombarding ions which was less than the thickness of the foil. We can therefore assume that Ar-ions implanted in the foil reduce or even prevent the effect of the displacements and hold the foil intact.

The lifetime of our standard foils was now compared with that of foils which were bombarded for 7 min. with 10 keV Ar-ions of density $\sim$ 20 $\mu A/cm^2$. Under exactly similar conditions, viz., using focused 10 keV He-ions of density $\sim$ 100 $\mu A/cm^2$, the average value of the lifetime in this case was measured to be $\tau \simeq 15$ min.

Conclusions

The experiments have clearly shown that carbon foils bombarded by low energy Ar-ions have lifetimes five times longer than that of our standard foils. We have no evidence that this effect is due to structural changes in the foil, but it can be attributed to the argon implanted during irradiation. As proof of the Ar-content of the foils, PIXE-analysis yielded an atomic concentration of 1.7%. Further work is necessary to clarify whether an optimum value exists for the concentration of the implanted atoms. Presumably, a yet longer lifetime will be attained if the implantation is uniformly distributed throughout the thickness of the foil.

It is also feasible that the lifetime depends on the species of the implanted atoms and/or better results are achieved if the implantation takes place during the growth of the foil, i.e. during deposition. Finally, the question arises as to whether longer-lived foils are always obtained by the presence of a suitable contaminant (dopant) which could be introduced during the fabrication process.

REFERENCES

(1) J. P. Biersack, Z. Physik 211 (1968) 495.
(2) P. Dobberstein and L. Henke, Nucl. Instr. and Meth. 119 (1974) 611.
(3) A. E. Livingston, H. G. Berry, and G. E. Thomas, Nucl. Instr. and Meth. 148 (1978) 125.
(4) D. Balzer and G. Bonani, Proc. INTDS Conf. (Munich, 1978), Nucl. Instr. and Meth. 167 (1979) 129.

EXPERIENCE WITH FOIL STRIPPERS IN THE

CHALK RIVER MP TANDEM ACCELERATOR

J. L. Gallant, D. Yaraskavitch, N. Burn,
A. B. McDonald, and H. R. Andrews

Atomic Energy of Canada Limited
Physics Division
Chalk River Nuclear Laboratories
Chalk River, Ontario, Canada KOJ 1J0

ABSTRACT

Techniques are described for the preparation of stripper foils. Foils prepared using three of these techniques were tested in the terminal of the Chalk River MP Tandem Accelerator using heavy ion beams such as bromine and iodine. One technique produced foils with a useful lifetime 30 times longer than those previously in use. This paper will discuss recent developments in fabrication and the procedures used for evaluating the test results.

INTRODUCTION

Under ion beam bombardment stresses develop in carbon stripper foils due to radiation damage; the heavier the ion, the faster the destruction of the foil. When the Chalk River MP Tandem Accelerator went into operation in April 1967, only gas was used for ion stripping in the high voltage terminal. Gas stripping has the advantage of producing less energy straggling and less multiple scattering than foil stripping; the disadvantage of gas stripping is that it yields lower average charge states than foil stripping. As experiments moved towards heavier ions at higher energies, higher average charge states were required. As a result in 1972, a National Electrostatics Corporation foil changing mechanism was installed in the terminal as part of the upgrading program of the MP. Tests were made to determine foil lifetimes and currents obtainable with the newly installed foil stripper[1].

The carbon stripper foils used on a routine basis since then (2 and 5 $\mu g/cm^2$ thick) were prepared by subliming carbon with an electron gun(2). For ions below atomic number 20 the carbon foil lifetime was usually acceptable ($\gtrsim$ 1 hour). However, for heavier ions, the lifetimes were substantially shorter. Recently work at other laboratories(3,4,5) indicated that foil lifetimes could be enhanced by relieving the radiation induced stresses or by producing another carbon structure that would delay the structural damage phenomenon. To this effect, three techniques developed recently were tried alone and in combination. They are:

a) The evaporation of carbon from a carbon arc onto a substrate heated at 300°C(3).

b) Slackening of the carbon foil by reducing the diameter of the supporting ring(4).

c) Production of carbon films by high voltage cracking of ethylene gas(5).

Figure 1. The apparatus used to prepare JAERI type foils showing the carbon arcing mechanism, the nickel chloride evaporation source, the substrate heater and the thickness monitor.

The Evaporation of Carbon onto a Glass Substrate Heated to 300°C

The JAERI (Japan Atomic Energy Research Institute) method[3] consists of evaporating carbon from a carbon-arc discharge (Figure 1) onto a glass slide which is heated to 300°C and has previously been coated with a thin film of nickel chloride. The arc gap is between 1 and 3 mm. The authors of this method have performed an elaborate pre-treatment of the glass slides. However, the crucial aspect of the method appears to be the heating of the substrate to 300°C.

We have obtained satisfactory results simply by rinsing the glass slides in alcohol and fire polishing them.

The Slackening of Carbon Foils by Reducing the Diameter of the Supporting Ring

When carbon films under ion beam bombardment suffer radiation induced shrinkage the surface becomes mirror-like and the foil

Figure 2. Collet-type apparatus for compressing aluminum rings on which are mounted carbon films. Four collet-type dies are used in one operation to produce four slackened foils.

breaks shortly afterwards. If there could be movement of the carbon film on the surface of its holder, it would relieve the stresses and prolong the foil lifetime. However, tests with holder surfaces coated with teflon or diffusion pump oil showed no lifetime improvement. As an alternative the Daresbury Laboratory has developed a collet-type apparatus to shrink the diameter of an aluminum supporting ring by approx. 10%. This procedure relaxes the mounted foil by forming a dome-like structure which permits much more shrinkage before stress develops to the point of rupture. Our apparatus to shrink supporting rings is shown in Figure 2.

Rings 21 mm outside diameter, 16 mm inside diameter and 1.5 mm thick are machined from 57 S aluminum. To soften the aluminum, the rings are heated to ∿ 500°C in the hot zone of a gas flame then quenched in ethyl alcohol. The carbon films are mounted on the aluminum rings and dried. The rings are then placed in the collet-

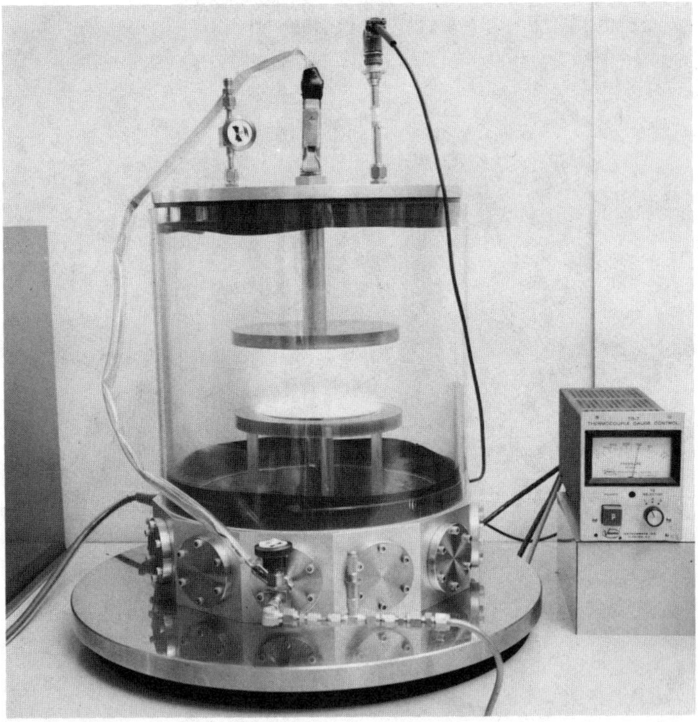

Figure 3. Ethylene cracking apparatus. The cathode sits on a tripod made of LEXAN (polycarbonate). All metal parts are covered with the plastic material so that the discharge will occur only between the two plates.

Figure 4. Apparatus for floating foils on distilled water. The mask for scribing the foils may be seen on the left.

Figure 5. Slackened carbon films made by the ethylene cracking method. At the far left the dome-type structure can be seen.

type slackening apparatus and using a hydraulic press, the outside diameter is reduced to 19 mm.

Production of Carbon Films by High Voltage Cracking of Ethylene Gas

This technique consists of dissociating ethylene, C_2H_4, in a high voltage field between two electrodes and collecting the carbon at the cathode. The cracking apparatus was simplified by using a 128 mm diameter Kindermann-type collector plate which sits on the cathode (Figure 3). In this procedure, the ethylene is cracked at a potential of 2.5 kV and a gas pressure of 10.7 Pa (80 microns of Hg). Carbon is collected onto the surface of the Kindermann plate which has previously been coated with a 10 $\mu g/cm^2$ film of sodium chloride.

The plate is then immersed in a dilute cellulose nitrate solution (100 ml collodion diluted to one liter with amyl acetate). The plate is dried vertically and the surface of the carbon scribed through a mask (Figure 4). The carbon is then floated on distilled water (Figure 4). The resulting films are mounted on aluminum rings, dried and pressed in the collet-type apparatus as previously described in The Slackening of Carbon Foils by Reducing the Diameter of the Supporting Ring. The relaxed ethylene cracked carbon films are shown in Figure 5.

Foil Lifetime Tests

Three methods of carbon film fabrication have been described. Initially, the slackening technique was applied to the standard carbon films routinely used in the terminal. The procedure was made easier by adding a thin cellulose film to the carbon surface as described in a previous paper(2). The slackening procedure resulted in a substantial improvement in the lifetime of these foils (see below). In addition, standard, JAERI type and cracked ethylene foils, some mounted on regular frames and others slackened, were evaluated (a) in an experimental beam line and (b) in the terminal of the MP accelerator.

Cracked ethylene and JAERI foils were also sent to Oak Ridge National Laboratory where they were compared with ORNL type carbon foils using a 10 MeV ^{35}Cl beam.

The carbon films tested on one of the experimental beam lines of the MP tandem were bombarded with a 1.5 mm diameter, 10.5 MeV $^{127}Iodine^{3+}$ beam. The total integrated ion beam current that caused the foils to break was measured in microcoulombs (μC). The foils evaluated under these conditions included the carbon films

Table 1. Foil Tests Performed at an Experimental Position with a 10.5 MeV Iodine Beam 3^+ Q. (μC Break)

AECL (1) EB (2) reg. 5 $\mu g/cm^2$	AECL EB reg. 10 $\mu g/cm^2$	AECL EB Slackened 5 $\mu g/cm^2$	AECL EB Slackened 10 $\mu g/cm^2$	JAERI arc (3) reg. 10 $\mu g/cm^2$	JAERI arc Slackened 10 $\mu g/cm^2$	Ethylene Cracking (4) reg. ~10 $\mu g/cm^2$	Ethylene Cracking Slackened ~10 $\mu g/cm^2$
60	60	853	853	355	3428	336	6617
60	60	853	1020	189	2859	866	6879
		1172	1172	350	1800	2377	2522
		523	1617	320	3000	2522	
		1072	1631	807	5600		
		908	1400	707	5600		
			1200	691	5600		
		481		825			
		961					
		961					
60	60	859	1270	530	3337	1427	5339

(1) Foils used routinely as stripper foils at ORNL.
(2) Method of Preparation: Electron bombardment.
(3) Method of Preparation: Arc discharge.
(4) Method of Preparation: High voltage cracking.

(5 and 10 μg/cm²) used on a routine basis. Half of these foils were slackened. JAERI foils and cracked ethylene foils (10 μg/cm²) were also tested as above.

Results

The results (Table 1) show that the foil slackening procedure enhances lifetimes and that the foils prepared by the new methods have a much greater life under bombardment. All foil types showed similar reactions to bombardment. The beam spot became mirror-like in appearance and wrinkles developed in the exterior regions. Breakage eventually resulted from this buildup of stress and therefore the initially slackened foils had longer lifetimes. The longer lifetime of the JAERI and ethylene type foils occurred because these foils developed stress at a lower rate. Elemental analysis of all three foil types by the scattering of ^{35}Cl ions(6) indicated an atomic percentage of hydrogen less than 3% in all three types, the remaining material being carbon. This strongly

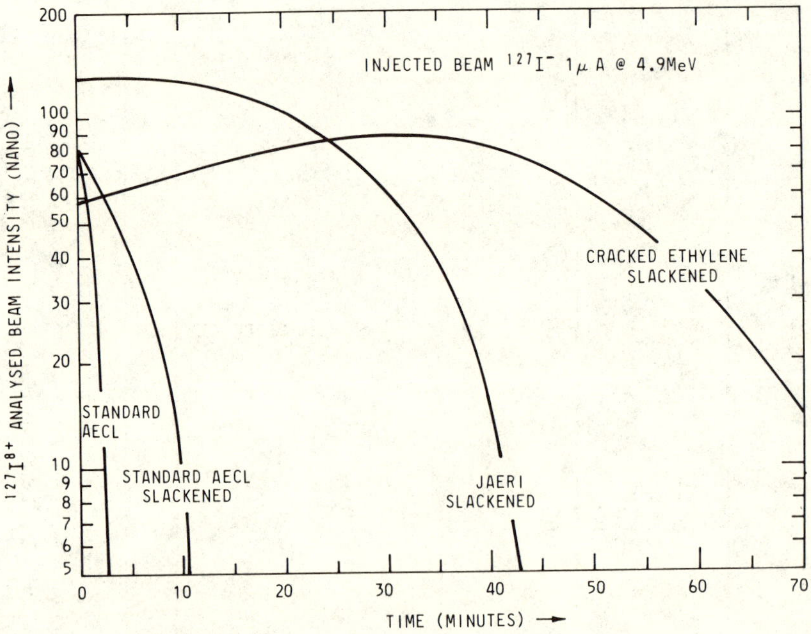

Figure 6. Typical variation of analyzed beam intensity with time for various foil strippers in the terminal of the MP Tandem accelerator.

Table 2. Foil Lifetime Tests Performed in the Terminal of the Chalk River Tandem Accelerator

(1 μA $^{127}I^-$ @ 4.9 MeV) 1979 March 14

FOIL LIFETIME TESTS

FOIL TYPE	NO. TESTED	LIFETIME IN MINUTES TO 1/2 MAXIMUM VALUE	LIFETIME IN MINUTES TO 1/10 MAXIMUM VALUE
Standard AECL	17	1.7 ± 0.4	4.0
Standard AECL (Slackened)	22	9.0 ± 0.8	20.0
JAERI (Slackened)	24	25.5 ± 4.1	50.0
Cracked Ethylene (Slackened)	8	47.7 ±20.2	73.0

Table 3. The Effect of Ion Mass and Intensity on Foil Lifetime at the Terminal of the Chalk River MP Tandem Accelerator

FOIL LIFETIME TESTS 1979 April 14

Standard AECL foils 2 μg/cm²-Slackened

Ion Type	Intensity (μA)	No. Tested	Lifetime in Minutes to 1/2 Maximum Value
$^{63}Cu^-$	1.2	35	112.6 ± 8.4
$^{63}Cu^-$	2.5	63	72.6 ± 6.0
$^{79}Br^-$	4.0	41	20.9 ± 2.3
*$^{127}I^-$	1.0	22	9.0 ± 0.8

*Included from 79 March 14 results for ease of comparison.

Table 4. Final Foil Lifetime Measurements at the Terminal of the Chalk River MP Tandem Accelerator

(1 μA $^{127}I^-$ @ 4.9 MeV) 1979 July 14

SLACKENED FOIL LIFETIME TESTS

Analyzed beam $^{127}I^{8+}$ 41 MeV

Foil Type	Max. Intensity (nA)	No. Tested	Lifetime in Minutes to 1/2 Maximum Value
Standard AECL	69.0 ± 5.3	5	6.2 ± 1.0
JAERI	*118.8 ± 8.9	18	27.2 ± 2.9
Cracked Ethylene	** 58.4 ± 5.3	15	44.9 ± 5.0

*Analyzed beam started off ∿ 10% below max. and reached this value in ∿ 10 minutes.
**Analyzed beam started off ∿ 20% below max. and reached this value in ∿ 20 minutes.

suggests that the difference in lifetimes must arise from different structure in the carbon, probably from different effective temperatures in the deposition process.

Tests at Oak Ridge(7) with 10 MeV ^{35}Cl also indicated lifetime improvements of ∿ 10 for unslackened ethylene and JAERI foils (made at ORNL) and > 40 for slackened foils made by both techniques when compared with standard vacuum deposition techniques.

The ultimate test in assessing the practical lifetime of a foil is to use it as a stripper in the terminal of an accelerator. Three series of tests were carried out with foils in the terminal of the MP tandem accelerator. In the first terminal trial, a 1 μA ^{127}Iodine beam was injected with the terminal at 4.9 MV. After a short initial period of constant intensity, or a slight intensity rise, analyzed beam intensities were observed to decrease steadily at a rate characteristic of the foil type (see Figure 6). This decrease is always observed and is thought to result from increased multiple scattering caused by a thickening of the foil. The results (Table 2) are expressed as foil lifetimes in minutes to 1/2 and to 1/10 of the maximum intensity of the analyzed beam. In the second series of terminal tests the effect on foil lifetime of ion beam mass and intensity was evaluated (Table 3). The purpose of the last series of tests (Table 4) was to confirm the earlier measurements (Table 2) but with improved statistics.

The difference in maximum intensity observed for the three foil types (Table 4) may be due to the difference in thickness of foil producing a variation in transmission due to multiple scattering. Accurate thickness measurements of the three foil types will be made after they are removed from the accelerator to determine if the apparent intensity advantage of JAERI foils is real.

Conclusion

The following conclusions can be drawn from the results shown:

1. Slackening of a carbon stripper foil will prolong its lifetime under ion beam bombardment.

2. Carbon film deposition under heated substrate conditions appears to be different from the cold substrate growth and results in longer foil lifetimes.

3. Carbon films prepared by cracking ethylene appear to have a different structure and one that is more resistant to radiation induced shrinkage than vacuum deposited films. The combination of the hot substrate or ethylene cracking techniques with the foil slackening method may produce films with an even greater practical lifetime.

It is now possible to use in the terminal of accelerators, carbon films with acceptable lifetimes for very heavy ion beams such as copper, bromine and iodine.

Compared to non-slackened standard foils, slackening of the standard foils gives a factor of $\sim$ 5 improvement in lifetime, slackening of JAERI foils gives a factor of 15 improvement and slackening of cracked ethylene foils gives a factor of 25 improvement.

ACKNOWLEDGEMENTS

The authors wish to thank Dr. Ronald L. Auble, Oak Ridge National Laboratory, for performing tests on AECL foils; D. Tolfree, Daresbury Laboratory for the exchange of information and J. P. Labrie and J. L'Ecuyer of the Université de Montréal for analyzing the hydrogen content of the foils.

REFERENCES

(1) Progress Report, Physics Division PR-P-96, Chalk River Nuclear Laboratories, Chalk River, Ontario AECL-4428 (p-18).

(2) Proceedings of the 1974 Annual Conference of the Nuclear Target Development Society, Chalk River Nuclear Laboratories, Chalk River, Ontario AECL-5503 (p-172).
(3) S. Takeuchi, C. Kobayashi, Y. Satoh, T. Yoshida, E. Takekoshi, and M. Maruyama, Nucl. Instr. and Methods 158 (1979) 333.
(4) D. W. L. Tolfree, D. S. Whitmell, and B. H. Armitage, Nucl. Instr. and Methods 163 (1979) 1.
(5) D. S. Whitmell, B. H. Armitage, D. W. L. Tolfree, and N. R. S. Tait, Daresbury Laboratory DL/NSF/P 86 (accelerator).
(6) J. P. Labrie, Université de Montréal, Private Communication.
(7) R. L. Auble, Private Communication.

HEAVY ION STRIPPING BY WRINKLED CARBON FOILS

P. K. Den Hartog, J. L. Yntema, G. E. Thomas,
and W. Henning

Argonne National Laboratory
Argonne, Illinois 60439

We report on measurements performed at Argonne to study the properties of wrinkled foils produced by the technique that was first developed at Daresbury(1). Foil lifetimes and transmission properties were measured under exposure to a 3 MeV ^{84}Kr beam at the Argonne Dynamitron accelerator, and to an 8 MeV ^{58}Ni beam at the terminal of the Argonne FN tandem. Energy loss straggling was measured by the time-of-flight method utilizing the picosecond beam bunching system of the Argonne superconducting heavy-ion linear accelerator.

A more durable carbon foil for ion beam stripping in tandem accelerators has been sought by many laboratories(2-11). Since foil stripping yields a higher average charge state than gas stripping, more energetic ions are produced, effectively increasing the useful energy range of the accelerator. As research at tandem accelerators is focused increasingly on nuclear physics with heavier ions, the lifetime limits of foils will become increasingly troublesome. Although the mechanism of foil destruction is not well known, foil lifetimes have been found to decrease with decreasing ion velocity(9).

The higher energies obtained through foil stripping are gained at the expense of accelerator operating time. At Argonne, we have frequently injected currents of 1 μA of ^{32}S at a terminal voltage of 8.5 MV. Under these conditions, foil lifetime is only about 1 hour. Our foil changer is a commercially available model* which holds 230 foils; yet this consumption rate implies in

*National Electrostatics Corp., Middleton, Wisconsin.

principle a new foil loading every ten days. In practice, due to down time associated with the experimental program, tank openings occur at 20 to 30 day intervals. Foil replacement is, in general, the primary reason for nearly all our tank openings. In the past, foil replacement required venting of the accelerator tubes and subsequent bakeout and conditioning. This time-consuming process has been greatly simplified in the last year by the installation of terminal valves similar to those designed and installed at Canberra[12]. Pump out, repressurization, and conditioning can now be accomplished in 36 hours or less. Still, at this foil consumption rate the mounting of foils is time consuming.

Although gas stripping does reduce the average charge state, it is still an attractive option for those experiments which do not require maximum energy. This is especially true for tandem

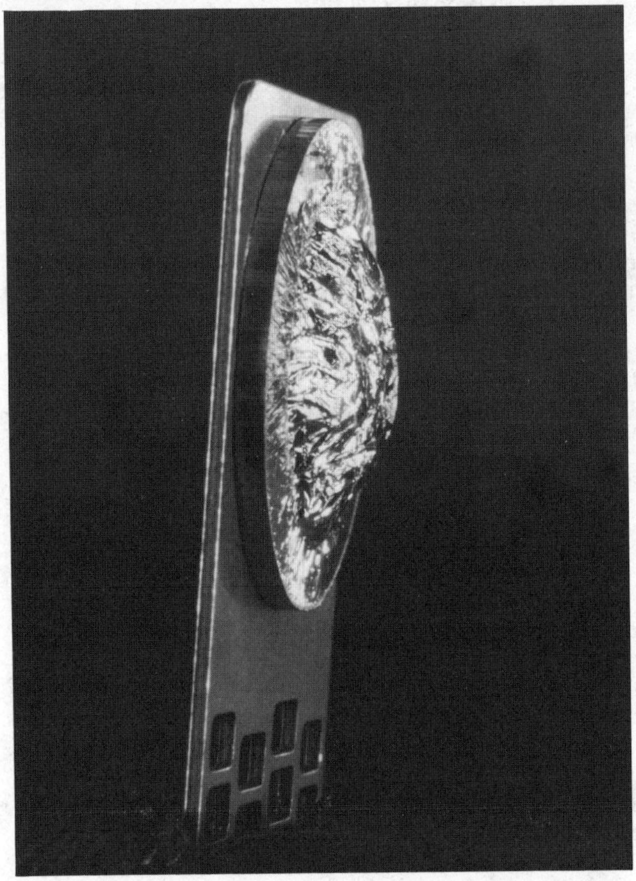

Figure 1. 5 $\mu g/cm^2$ wrinkled foil before irradiation.

accelerators combined with post-accelerators such as at Argonne. Nevertheless, gas stripping is difficult to accomplish in a vacuum upgraded tandem. Differential terminal pumping, with all its associated complexities, is most likely required. For all these reasons, we found the results which were reported from Daresbury(1) utilizing mechanically slackened or wrinkled foils very encouraging.

The wrinkled foils are attractive because of the straightforward method of extending the lifetime. Whatever the microscopic mechanism, clearly the immediate cause of foil breakage is the extreme stress developed within the foil. We have produced wrinkled foils by the Daresbury technique. Figure 1 shows the typical appearance of a wrinkled foil before irradiation. The initial aperture area of the foil has been reduced approximately 6%.

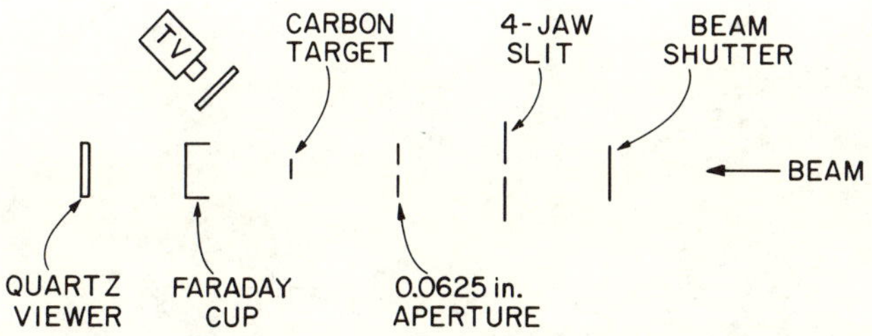

Figure 2. Beamline layout for foil lifetime measurement.

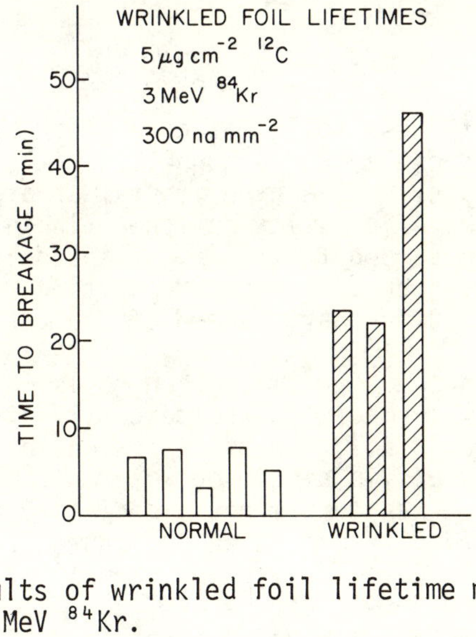

Figure 3. Results of wrinkled foil lifetime measurements using 3.0 MeV ^{84}Kr.

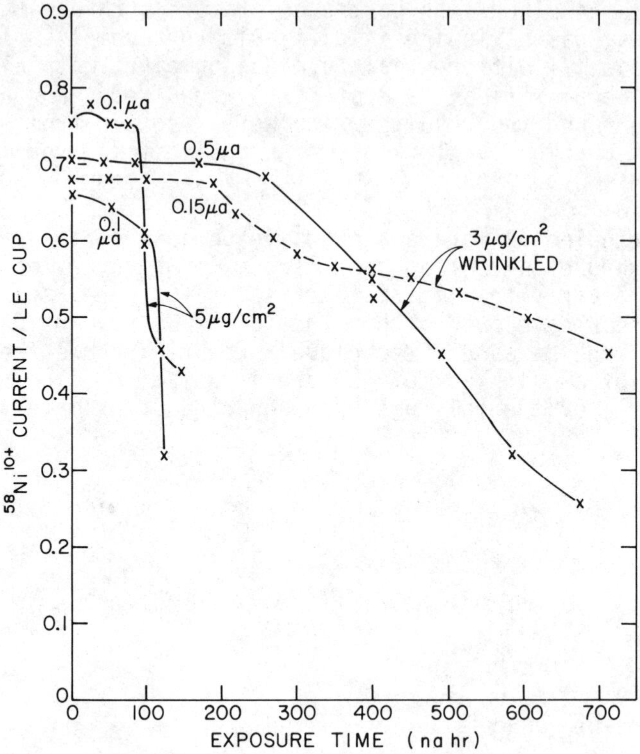

Figure 4. $^{58}Ni^{10+}$ transmitted through Argonne FN Tandem Accelerator using normal and wrinkled carbon foil strippers.

The lifetime of these foils was determined by exposure to a 3 MeV ^{84}Kr beam produced by the Argonne Dynamitron accelerator. Figure 2 shows a layout of the experimental apparatus. The beam was collimated by the 4-jaw slits and the 1/16 inch diameter aperture in front of the carbon foil. The stripped ion beam was collected in a faraday cup. The foil condition was monitored by the TV camera and a strip chart record of the current.

Normal and wrinkled 5 μg/cm² carbon foils were alternately tested. All foils were made from commercially available slides*. Each was dipped in a 7% solution of collodion in isoamyl acetate before being floated off in distilled water and mounted. The incident beam current was determined immediately prior to introducing the foil. The results are shown in Figure 3. The normal foils had a relatively well defined lifetime, averaging about 6 minutes

*Arizona Carbon Foil Co., Tucson, Arizona.

to breakage. The wrinkled foils survived for an average of 30 minutes although one lasted considerably longer than the others.

Offline tests are a useful test of stripper performance, but the final test must be done in the terminal of an accelerator. In most offline tests, current densities are higher and lifetimes shorter than for foils exposed in a terminal. Also, most offline tests reveal only the total beam transmitted while in practice the transmission for a particular charge state is of more interest.

The 3 μg/cm^2 and 5 μg/cm^2 wrinkled foils were loaded into the terminal of the Argonne FN tandem along with 5 μg/cm^2 normal foils. The foils were irradiated at a terminal voltage of 8 MV by beams of 150 nA and 500 nA of $^{58}Ni^-$, as measured at the low energy cup. The results are shown in Figure 4 where the transmission ratio of $^{58}Ni^{10+}$ (i.e., the ratio of the analyzed $^{58}Ni^{10+}$ charge current to the total beam current at the low energy faraday cup) has been plotted against the product of exposure time and beam current in nA hours. This transmission ratio depends on the characteristics of the accelerator and on the multiple scattering of the incident beam by the foil.

The transmission ratio for the normal foils remains essentially constant until breakage at about 125 nA hours. The wrinkled foils, however, did not show any precipitous fall in transmitted beam, even after 600 nA hours. They do show a steady decrease in the transmission ratio, probably as a result of foil thickening. The initial transmission ratio for 3 μg/cm^2 wrinkled foils is 0.7, which is about equal to the initial transmission ratio of a 5 μg/cm^2 normal foil. The initial ratio for a 5 μg/cm^2 wrinkled foil is between 0.5 and 0.55.

The accelerator and terminal form a baked all-metal ceramic structure which is pumped at both ends by ion and cryopumps. The pressure at the base of the accelerator is typically in the 10^{-8} torr range, and in the non-pressurized machine the terminal pressure, of course without beam, is typically in the low 10^{-7} torr range. Nevertheless there is some indication that the apparent foil thickening process is related to the partial pressure of hydrocarbons in the terminal. We observe that the gas pressure in the accelerator system shows increases proportional to the heavy ion beam current and increases with the mass of the accelerated ion. One possible source of the hydrocarbons is beam induced outgassing of defining apertures. Ions and electrons leaving the foil outside of the acceptance angle of the aperture may liberate hydrocarbons from the metallic surfaces. An additional effect contributing to the thickening may be migration of foil material into the region of the beam spot from the remainder of

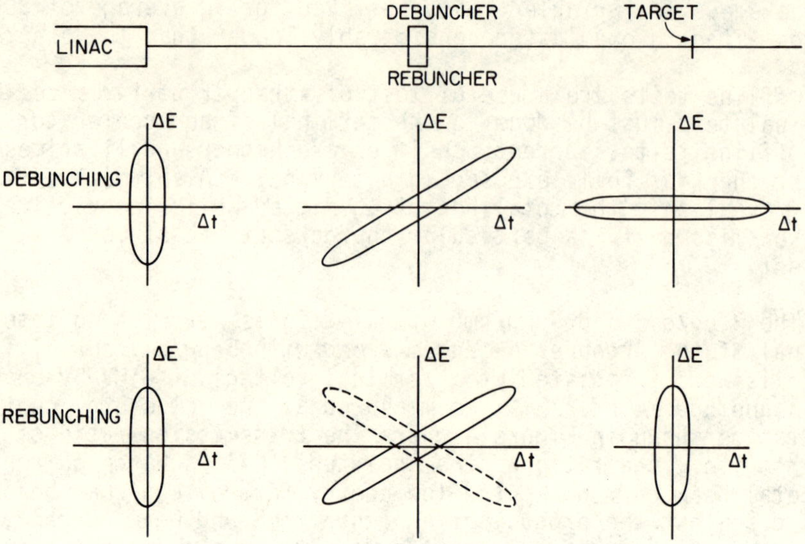

Figure 5. The effect of rebunching or debunching on longitudinal phase space ellipse after linac exit.

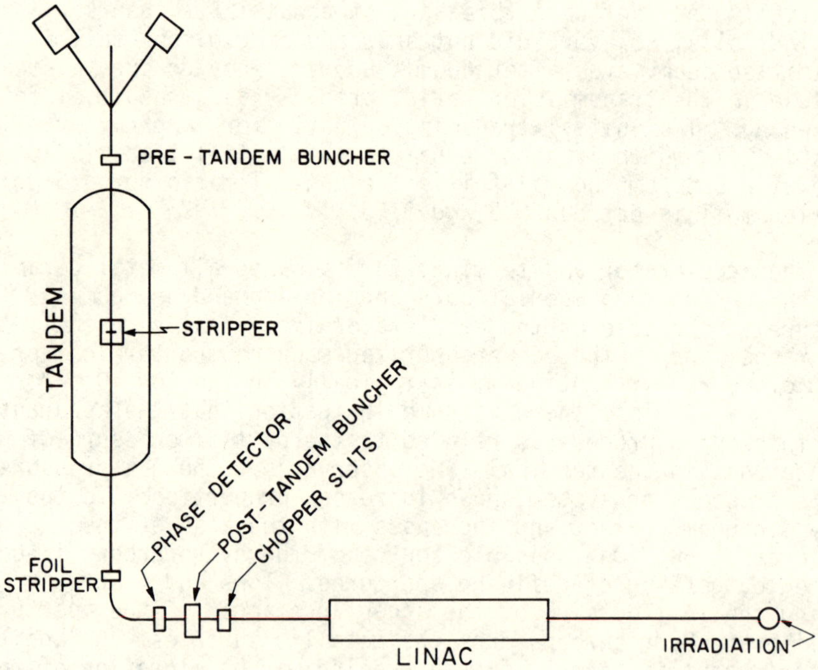

Figure 6. Layout of system used for measurement of energy loss straggling.

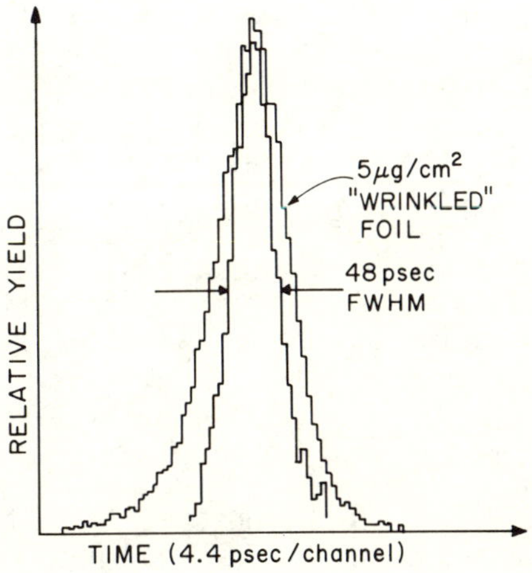

Figure 7. Time spectra of beam using system of Figure 6 described in text. The 48 psec FWHM peak is the system resolution. The beam was 50 MeV ^{16}O.

the foil. The future addition of a terminal ion pump should help in determining the relative contributions of these two effects.

No new stripping medium proposed for use at the Argonne tandem-linac can be accepted on the basis of lifetime alone. Of equal importance is the energy spread introduced into the beam by energy loss straggling. The quality of experimental data obtained with the tandem-linac depends in many instances on the longitudinal emittance, $\Delta E \Delta t$, of the beam[13]. This is schematically demonstrated in Figure 5 in terms of the evolution of the longitudinal phase space ellipse. Depending on the need for good timing or good energy resolution, the beam as it leaves the linac may be rebunched or debunched in the manner illustrated in Figure 5. The energy loss straggling introduced by the stripping foil effects the area of the phase ellipse, irrevocably degrading the beam quality.

The energy straggling of carbon foils was measured by time-of-flight using the pico-second beam bunching system shown in Figure 6[14]. The system consists of a gridded harmonic pre-buncher, chopper, phase detector, a superconducting post-tandem buncher, and an effective flight path of approximately 20 m. To measure the intrinsic time resolution of the system, the last resonator in the linac was used as a rebuncher providing for a short flight path and consequently reduced time spread. The measured time width for ^{16}O was 48 psec. Correcting for the time spread due to the 4 m drift length from the last resonator to the detector, the system time resolution is determined to be 40 psec.

The time spectrum demonstrating the system resolution is shown in Figure 7, together with a spectrum obtained for a 5 μg/cm^2 wrinkled foil for ^{16}O for the full flight path. The time width for the wrinkled foil corresponds approximately to that measured for a 8 μg/cm^2 normal foil. The final results indicate that the energy broadening in arc-evaporated carbon foils varies approximately as $t^{1/2}$ where t is the thickness of the foil, as one would expect if energy loss straggling is the dominant contribution to the energy width. No significant tails of energy distribution anomalies are found which would make the use of wrinkled foils unacceptable.

REFERENCES

(1) B. H. Armitage, J. D. H. Hughes, D. S. Whitmell, N. R. S. Tait, and D. W. L. Tolfree, Nucl. Instr. and Meth. 155 (1978) 565.

(2) J. L. Yntema, Nucl. Instr. and Meth. 122 (1974) 45, ibid 113 (1973) 605, and ibid 98 (1972) 379.

(3) D. S. Whitmell, B. H. Armitage, D. R. Porter, and A. T. G. Ferguson, Proc. Int. Conf. on the Technology of Electrostatic Accelerators, Daresbury, 1973, DNPL/NSF/R5 (1973) 265.

(4) P. Dobberstein and L. Henke, Nucl. Instr. and Meth. 119 (1974) 611.

(5) P. D. Dumont, A. E. Livingston, Y. Boushnet-Robinet, G. Weber, and L. Quaglia, Phys, Sci. 13 (1976) 122.

(6) J. L. Yntema, IEEE Trans. Nucl. Sci. NS-23, No. 2 (1976) 1133.

(7) G. E. Thomas, P. K. Den Hartog, J. J. Bicek, and J. L. Yntema, Proc. 6th Annual Conference of the INTDS, Berkeley, California, 19-21 October, 1977, LBL 7950.

(8) G. E. Thomas, P. K. Den Hartog, J. L. Yntema, and R. D. McKeown, Proc. 7th Annual Conference of the INTDS, Garching, Germany, 11-14 September, 1978, Nucl. Instr. and Meth. 167 (1979) 29.

(9) A. E. Livingston, M. G. Berry, and G. E. Thomas, Nucl. Instr. and Meth. 198 (1978) 125.

(10) K. R. Chapman, Nucl. Instr. and Meth. 198 (1978) 209.

(11) S. Takeushi, C. Kobayashi, Y. Satoh, T. Yoshida, E. Takekoshi, E. Takekoshi, and M. Maruyama, Nucl. Instr. and Meth. 158 (1979) 333.

(12) The drawings of the Canberra terminal valves were provided by Dr. D. Weiser.

(13) J. Aron, R. Benaroya, L. M. Bollinger, B. G. Clifft, W. Henning, K. W. Johnson, J. M. Nixon, P. Markovich, and K. W. Shepard, Proc. of the 1979 Linear Accelerator Conference; and references therein.

(14) F. J. Lynch, R. N. Lewis, L. M. Bollinger, W. Henning, and O. D. Despe, Nucl. Instr. and Meth. 159 (1979) 245.

A REVIEW OF METHODS TO PREPARE SELF-SUPPORTING CARBON TARGETS AND OF THEIR IMPORTANCE IN ACCELERATOR EXPERIMENTS

P. Maier-Komor

Physik-Department
Technische Universität München
D-8046 Garching, West Germany

I. INTRODUCTION

Carbon occupies a special position compared to all other elements. Without carbon, the basis for life would be impossible. Organic chemistry or carbon chemistry is an independent part of chemistry. Just so the element carbon occupies an exceptional position in target preparation for nuclear measurements due to its melting point, the highest of all elements, the very low vapor pressure and the chemical stability which is comparable to that of the noble metals. Carbon has only two stable isotopes and one radioactive isotope the half-life of which is long enough to prepare targets. The natural abundance of carbon-12 is nearly 99% which means a natural carbon target can be used in most cases as isotopic ^{12}C-target. But this is only one part of the story, the other one is given by the excellent properties of the self-supporting foils. The good tensile strength of an amorphous carbon layer can be estimated by the minimum thickness of a self-supporting foil. Carbon foils can be prepared even below 1 $\mu g/cm^2$ ($\approx$ 60 Å), they are one order of magnitude thinner than foils of any other element.

The use of carbon foils is unavoidable in nearly all experiments on nuclear physics and related subjects performed at ion accelerators.

The ion energy necessary to perform the wanted nuclear reaction is too high to be achieved by acceleration of a single charged particle.

Mostly carbon foils are installed in the beam line of the accelerator to increase the charge state of the accelerated ions. Due to the limited lifetime of these carbon foils in the ion beams, the accelerators are the biggest users of carbon foils.

Most accelerator experiments carried out at thin targets below 100 $\mu g/cm^2$ have to live with the target on a thin backing material and this is in most cases a carbon backing in a thickness range of 3 - 30 $\mu g/cm^2$. There are also whole branches of accelerator physics, where mostly carbon foils are used as targets. One example is the beam foil spectroscopy, which leads to information about the interaction of the atoms in the foil with the penetrating ions and the decay of the excited states of the ions by looking at the emitted light behind a carbon foil. Another example is "Physics with Fast Molecular Ion Beams" established by Gemmel(1): amorphous carbon foils are bombarded with molecular-ion beams. Due to stripping in the carbon foil a molecular ion dissociates into its fragments, their energies and angular distributions are measured to get information about the structure of the molecule.

II. METHODS OF FABRICATION

A. Mechanical Techniques

Mechanical preparation methods are successful down to a minimum carbon foil thickness of several mg/cm^2. Hot pressed carbon or polycrystalline graphite material can be grinded and polished to a minimum thickness of 20 mg/cm^2. Single crystals of graphite can be cleaved, the minimum self-supporting thickness is about 2 mg/cm^2(2). Diamonds may be cut and polished down to thicknesses of 50 mg/cm^2(3). Graphite powder can be settled and pressed to give carbon foils of 1 mg/cm^2 minimum thickness(4). Carbon foils can also be prepared using a solution in which graphite powder is dispersed(5). This liquid is used to immerse and withdraw e.g. microscope slides, which are coated with a parting agent to float the carbon foils. Some of these suspensions are utilized to paint(6) or spray the carbon material on a parting agent coated substrate. Carbon foils of such kind can be prepared from 10 $\mu g/cm^2$ upwards.

B. Decomposition of Organic Compounds

Two different methods are established

Nowadays thermal cracking is utilized only to prepare ^{14}C-targets.

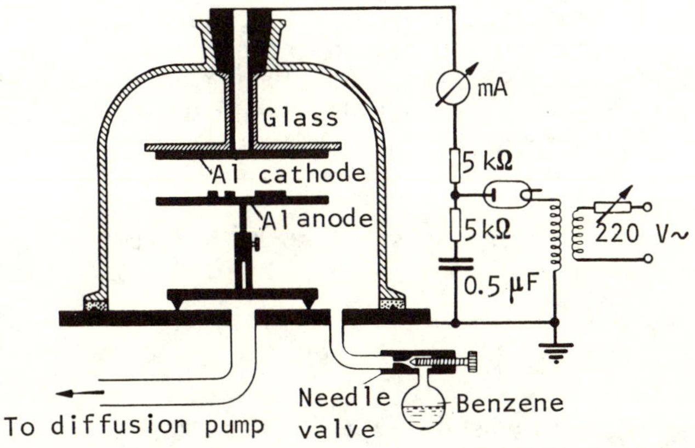

Figure 1. Glow discharge apparatus used by König and Helwig to crack hydrocarbon gases.

In principle a metal foil filament is heated in a vacuum chamber backfilled with a vapor of a suitable organic compound. The vapor molecules may be decomposed at the surface of the hot filament, but only carbon the element of low vapor pressure is deposited at the filament. The filament material can be etched away in an acid bath to get the self-supporting carbon foil.

In practice this method was only successful using methyl iodide as C-compound and a nickel filament(7-11).

Recently H. J. Maier(12) has improved this method. He exchanged the glass apparatus with greased stop cocks for a stainless steel construction to prevent the poisoning of the methyl iodide caused by its attacks of all kinds of grease. He presented precise data on the pressure of the methyl iodide and the temperature of the nickel filament. With such an apparatus it is possible to reproduce the thickness, the chemical purity and mechanical stability of such carbon foils. The carbon foil thickness ranges from 4 $\mu g/cm^2$ to 100 $\mu g/cm^2$.

The second method is the cracking of hydrocarbon gases in a DC-glow discharge chamber. This method is very popular at the moment, because Tait and Tolfree(13) discovered, that carbon foils - prepared by this method - live longer under heavy ion bombardment than foils prepared by the evaporation condensation process.

König and Helwig(14) developed this method in 1950.

Figure 1 shows a drawing of their glow discharge apparatus. It consists of a glass vacuum chamber containing a metal anode cathode electrode structure. Pyrex glass slides coated by a suitable parting agent are mounted on one electrode. A 2 kV power supply is connected with the electrodes via a vacuum feedthrough. After the chamber had been evacuated by a diffusion pump the hydrocarbon gas was leaked into the chamber to strike the glow discharge.

The minimum foil thickness seems to be 5 $\mu g/cm^2$ in the moment.

C. Sputtering

Sputtering is mainly used for all industrial applications where a thin film of good adherence to the substrate is needed. In the physical sputtering process the impact energy of the sputtered atom or molecule is several orders higher than the thermal energy brought in by an evaporation process. Because of the resultant good adherence to the substrate no reliable way of preparing self-supporting foils is reported in all the papers describing the sputtering process of carbon. Wirth(15) used the focused ion beam sputtering method first reported by Sletten(16) to prepare carbon films on rock salt crystals. But the very low efficiency during the floating process let him give up this procedure.

D. Evaporation Condensation Process in High Vacuum

The evaporation of carbon is carried out by three different methods: a) Resistance heating of carbon filaments, b) Arc evaporation, c) Electron bombardment heating.

The resistance heating method is used most widely, but the method published first by Bradley(17) is modified by many laboratories. Bradley described his method as carbon-arc evaporation despite the fact that the two electrodes are always in physical contact. Two pointed carbon rods are pressed together horizontally by springs. The points of the carbon rods evaporate if a current of more than 100 A passes through the contact point. Dearnaley(18), Sarma(19), and McCormick(20) used similar set ups. Nobes(21) modified this method by mounting the carbon rods vertically and by exchanging the spring loading by gravity using a weight of 80 g. The disadvantage of the methods above described is, that the carbon rods must be sharpened several times to evaporate enough carbon for thicker films, which requires braking of the vacuum. Maier-Komor(22) tried to avoid this inconvenience by resistance heating of a 3 mm diam. carbon rod mounted in fitting boreholes of spring-loaded carbon rod electrodes.

Arc evaporation was published first by Pfeiffer(23) and Blue(24), but both authors didn't see a difference to the Bradley method. The arrangement of the carbon evaporator is indeed the same as that of the Bradley method. In addition there must be a push-pull feedthrough connected to one electrode to adjust the arc gap from outside the vacuum system. Takeuchi et al(25) used a power supply of an argon-arc welder with a no-load voltage of 80 V. After touching the electrodes the carbon plasma burnt in the arc gap of 1-3 mm and the voltage dropped to about 20 V.

In the Bradley method the condensation energy of the carbon atoms is determined by a thermal process. The highest possible evaporation energy is given by a temperature of 2950 K(26), above which the evaporation is accompanied by a significant particle emission. In the plasma of the carbon-arc however, the carbon atoms can gain much more energy than in a pure thermal process. Thus the condensation mechanism can occur differently, which may result in a different structure of the carbon film.

Since the first publications of carbon evaporation by electron bombardment(27,28) this method has become very popular, because the evaporation rate control is very easy and the carbon source is nearly unlimited. You can use a thick long carbon rod, which is fed vertically into the electron gun crucible.

The evaporation of carbon in a high vacuum system is the commonly used method to prepare carbon films. A successful floating procedure of these films depends mainly on the parting agent and its coating on the substrates, which are glass microscope slides in most cases.

The crystalline structure of the parting agent does not influence the growth of the carbon layer, because all useful substrate temperatures are too cold compared to the absolute melting point of carbon. So the carbon will always condense amorphous. The mechanical durability of evaporated carbon foils depends mostly on the surface roughness of the parting agent, especially for those thin foils used as tandem terminal strippers. Standard carbon foils grown on Betaine(22) with a thickness of 3 $\mu g/cm^2$ can easily be installed in the foil changing mechanism, whereas foils of the same thickness grown on a smooth surface of a parting agent need a collodion backing(29) to be stable enough for being mounted. Braski showed carbon replicas of several commonly used parting agents under different evaporation conditions(30). His measured surface roughnesses of the carbon foils permit conclusions on the homogeneity of the foils. Abele demonstrated that this non-uniformity caused by the parting agent influences strongly the energy loss straggling(31,32). This fact should not be neglected, when choosing a parting agent for a backing used in a high resolution experiment or for a stripper foil used in a tandem which should let the beam pulses keep their sharp energy and time.

Most laboratories use detergents as parting agents. The carefully cleaned substrates are coated with a thin layer of the liquid detergent by means of a Kleenex tissue. The surface is polished as it dries so that only a thin invisible layer remains. Some authors describe another way: Microscope slides are dipped into an aqueous detergent solution and dried in a vertical position. A base material for many industrial detergents is Teepol which is identical with Lensodel, this product is fabricated by Shell Chemical. The purity of Teepol guarantees that no other impurities than of the Teepol compound itself can be detected in the carbon foil. Other used detergents like RBS-25, RBS-50, Knox-60, PCB-100, and Creme-Cote, etc. are mixtures of different chemical compounds to get special cleaning effects.

Creme-Cote e.g. seems even to be perfumed. So the impurities in the carbon foils may be higher and not reproducible. Another group of parting agents in use are sugars. The aqueous solution of Betaine or Sucrose, or a mixture of both(22), is wiped with a fuzz-free tissue onto one surface of the glass slide, the wiping is proceeded until a dry homogeneous milky film is achieved. Carbon foils produced on such a parting agent have a very high mechanical stability, e.g. a target frame covered with a carbon foil, even thinner than 5 $\mu g/cm^2$, can be bent in the center around an angle of close to 180° without rupturing the carbon foil. This is due to the corrugated structure(32) of the foil. Misgivings, that a foil of this inhomogeneous structure causes more energy loss straggling than one of a smooth surface, are unreasonable. The energy loss straggling is equal for both kinds of foils as long as the incoming and outgoing beam is nearly perpendicular to the foil surface.

The third group of parting agents for carbon films are the halides, in most cases the chlorides. The carefully cleaned microscope slides are coated in a high vacuum system by an evaporation condensation process. These parting agents are used for the production of carbon foils in cases where a close source to substrate distance is essential, because heat radiation would destroy an organic parting agent.

III. CARBON STRIPPER FOILS

Carbon foil preparation seemed to be no longer a problem a few years ago. Now it has become a very interesting topic again, because the commonly used standard carbon stripper foils cannot satisfy the demands of heavy ion experiments which are numerously performed in tandem accelerators nowadays.

In 1972 Yntema directed the target makers to the problem of short lifetimes of carbon stripper foils penetrated by heavy ions of high energy[33]. Dobberstein[34] and Livingston[35] showed how crucial the problem of stripping heavy ions is, by presenting a lifetime drop of more than four decimal powers from the light to the heaviest atomic ions. After these papers many target laboratories started the development of a new generation of stripper foils. Of course there were enough proposals to prolong the lifetime of stripper foils by annealing or rotating the foil[36,37] during the beam penetration, but the methods need additional power supplies and driving mechanisms in the terminal of a tandem and in the end the prolongation of lifetime is not even satisfactory. The principle mechanism which causes the rupture of a stripper foil during ion bombardment is not well understood. The heat deposited in the foil due to the traversing heavy ion causes a transformation of the amorphous to a graphitized structure in the beam spot area. During this graphitization process the mass density is increased, which is accompanied by mechanical stress between the non-irradiated part and the beam spot area of the foil[38]. The rupturing is caused by this mechanical inhomogeneity.

This is the only well understood part of the story about foil rupture. The high energetic ions traversing the solid state stripper can damage the crystal structure in a catastrophic way not comparable to the smooth change caused by high temperatures. Doses of argon ions three times as many as the carbon atoms are necessary to rupture a carbon foil[25]. Who might believe that any carbon atom could maintain its position under the bombardment of a mass ten times its own one.

At the moment two principle ways are practicable in order to prolong the lifetime of carbon stripper foils. Both ways are initiated by the graphitization which takes place in the beam spot area of the foil during ion bombardment.

Tait[13] presented first foils of an amorphous structure that resist to this graphitization process for a longer time. These foils are prepared by the cracking of ethylene in a DC-glow discharge as described above. Balzer[39] implanted Ar-atoms in amorphous carbon foils to hinder the graphitization process. Maier-Komor[40] suggested first to prepare foils exhibiting already crystalline structure before exposure to the beam. In such foils the stress between the irradiated and the non-irradiated area can be prevented, because both parts have graphitized structure. The graphitization of the stripper foils was performed by Sander[41] and Maier-Komor[42] using different annealing methods. Sander heat-treated foils in an argon flushed oven at temperatures up to 1300 K. Maier-Komor heated the foils by electron bombardment in a high vacuum system up to 2200 K.

The method of Tait is repeated and improved nearly world-wide. Gallant[43] and Auble[44] could produce thinner ethylene cracked carbon foils by using the collodion coating procedure[29]. Meens[45] prepared thinner and probably more homogeneous ethylene cracked carbon foils using thin copper backings mounted on the cathode of the glow discharge apparatus. The floating procedure is done in an acid bath to etch away the copper foil. Huck[46] used Betaine as parting agent, which was protected against the thermal energies liberated by the glow discharge process. This protective coating was a thin vapor deposited NaCl layer. Carbon foils of such kind are less brittle and a collodion backing is not necessary even for foils below 5 $\mu g/cm^2$.

The aim of all these stripper foil developments should be to satisfy the tandem experimenters, who are only interested in the quantity of monoenergetic heavy ions per stripper foil available for their experiments. Carbon foils of 2-3 $\mu g/cm^2$ used as a terminal stripper give the best emittance for heavy ions analyzed in the most probable charge state.

This statement seems to be correct for all atomic ions used in tandem accelerators with voltages up to 15 MV. The new tandem generation with voltages up to 30 MV will probably demand carbon stripper foils below 6 $\mu g/cm^2$. Only lifetime data are interesting, which were measured on carbon foils of 2-6 $\mu g/cm^2$. These data must be presented in consideration of the analyzed beam[42]. Auble[44] has shown that in the case of the ethylene glow discharge foils the lifetime decreases to one tenth for a 2 $\mu g/cm^2$ foil compared to a 10 $\mu g/cm^2$ one.

These results show that a lot of development work has to be done until it will be evident that glow discharge carbon foils are long-living stripper foils. The other methods described above to prolong the stripper lifetime seem to be only at the beginning of their developments, too.

If all these methods will be perfectly developed then only the part of the destruction process is solved which concerns the graphitization during the heavy ion irradiation. Only when the whole destruction procedure of the stripper foils will be better understood some day, can we decide whether there is a possibility to fight against the other radiation damage effects by variations of the foil preparation methods, too.

REFERENCES

(1) D. S. Gemmel, E. P. Kanter, and W. J. Pietsch, Chem. Phys. Lett. 55 (1978) 331.
(2) G. E. Myers and G. L. Montet, J. Appl. Phys. 37 (1966) 4195.

(3) M. Rebak, J. P. F. Sellshop, T. E. Derry, and R. W. Fearick, Nucl. Instr. and Meth. 167 (1979) 115.
(4) W. R. Lozowski, Proc. 6th Ann. Conf. INTDS, Berkeley (LBL-7950) (1977) 115.
(5) V. E. Viola, Jr. and D. J. O'Connell, Nucl. Instr. and Meth. 32 (1965) 125.
(6) R. D. McCormick and J. D. McCormack, Nucl. Instr. and Meth. 13 (1961) 151.
(7) G. C. Phillips and J. E. Richardson, Rev. Sci. Instr. 21 (1950) 885.
(8) H. D. Holmgren, J. M. Blair, K. F. Famularo, T. F. Stratton, and R. V. Stuart, Rev. Sci. Instr. 25 (1954) 1026.
(9) E. Kashy, R. R. Perry, and J. R. Risser, Nucl. Instr. and Meth. 4 (1959) 167.
(10) A. H. F. Muggleton, Proc. Seminar on the Prep. and Standardisation of Isotopic Targets and Foils, Harwell, Oxon AERE-R5097 (1965) 99.
(11) R. Keller and H. H. Müller, Nucl. Instr. and Meth. 119 (1974) 321.
(12) H. J. Maier, these Proceedings.
(13) N. R. S. Tait, D. W. L. Tolfree, B. H. Armitage, and D. S. Whitmell, Nucl. Instr. and Meth. 167 (1979) 21.
(14) H. König and G. Helwig, Zeitschrift für Physik 129 (1951) 491.
(15) H. Baumann and H. L. Wirth, Nucl. Instr. and Meth. 167 (1979) 71.
(16) G. Sletten and P. Knudsen, Nucl. Instr. and Meth. 102 (1972) 459.
(17) D. E. Bradley, Brit. J. Appl. Phys. 5 (1954) 65.
(18) G. Dearnaley, Rev. Sci. Instr. 31 (1960) 197.
(19) N. Sarma, Nucl. Instr. 2 (1958) 361.
(20) R. D. McCormick and J. D. McCormack, Nucl. Instr. and Meth. 13 (1961) 147.
(21) M. Nobes, J. Sci. Instr. 42 (1965) 753.
(22) P. Maier-Komor, Nucl. Instr. and Meth. 102 (1972) 485.
(23) I. Pfeiffer, Naturwissenschaften 42 (1955) 508.
(24) M. D. Blue and G. C. Danielson, J. Appl. Phys. 28 (1957) 583.
(25) S. Takeuchi, C. Kobayashi, Y. Satoh, T. Yoshida, E. Takekoshi, and M. Maruyama, Nucl. Instr. and Meth. 158 (1979) 333.
(26) A. Greenville-Whittaker and P. Kinter, Carbon 7 (1969) 414.
(27) S. H. Maxman, Nucl. Instr. and Meth. 50 (1967) 53.
(28) M. Morgan, Thin Solid Films 7 (1971) 313.
(29) J. L. Gallant, Proc. 3rd Ann. Conf. of the INTDS, Chalk River, AECL-5503 (1974) 172.
(30) D. N. Braski, Nucl. Instr. and Meth. 102 (1972) 553.
(31) H. K. Abele, P. Glässel, P. Maier-Komor, H. Rösler, H. J. Scheerer, and H. Vonach, Proc. 4th Ann. Conf. INTDS, Argonne 1975, ANL/PHY/MSD-76-1, p. 117.

(32) H. K. Abele, P. Glässel, P. Maier-Komor, H. J. Scheerer, H. Rösler, and H. Vonach, Nucl. Instr. and Meth. 137 (1976) 157.
(33) J. L. Yntema, Nucl. Instr. and Meth. 98 (1972) 379.
(34) P. Dobberstein and L. Henke, Nucl. Instr. and Meth. 119 (1974) 611.
(35) A. E. Livingston, H. G. Berry, and G. E. Thomas, Nucl. Instr. and Meth. 148 (1978) 125.
(36) J. L. Yntema, Nucl. Instr. and Meth. 113 (1973) 605.
(37) G. E. Thomas, P. K. Den Hartog, J. L. Bicek, and J. L. Yntema, Proc. 6th Ann. Conf. INTDS, Berkeley 1977, LBL-7950.
(38) U. Sander, H. H. Bukow, and H. v. Buttlar, J. de Physique 40 (1979) C1-301.
(39) D. Balzer, these Proceedings.
(40) P. Maier-Komor, Proc. 5th Ann. Conf. INTDS, Los Alamos 1976, LA-6850-C.
(41) U. Sander and H. H. Bukow, Radiation Effects 40 (1979) 143.
(42) P. Maier-Komor and E. Ranzinger, these Proceedings.
(43) J. L. Gallant, D. Yaraskavitch, N. Burn, A. B. McDonald, and H. R. Andrews, these Proceedings.
(44) R. L. Auble and D. M. Galbraith, these Proceedings.
(45) A. Meens, Centre de Recherches Nucleaires, Strasbourg, France, private communication.
(46) B. Huck and H. Wirth, Max-Planck-Institut für Kernphysik, Heidelberg, West Germany, private communication.

GRAPHITIZATION OF CARBON STRIPPER FOILS

P. Maier-Komor and E. Ranzinger

Physik-Department, Technische Universität München
D-8046 Garching, West Germany

ABSTRACT

Procedures for increasing the lifetime of carbon stripper foils are described. Four to eight μg/cm² self supporting carbon foils were treated by two different methods in order to change their amorphous composition into a more crystalline structure. One method is the bombardment with electrons of energies up to 5 keV. In the second method both sides of the carbon foils are bombarded simultaneously with He^+ ions with energies up to 2 keV.

Both procedures result in a prolongation of the lifetime of the carbon foils used as terminal strippers in the Munich tandem accelerator. Results for S^- - beams of 9.0 and 10.6 MeV energy and intensities up to 2.5 μA are reported.

The improvement in lifetime can be explained by the crystal structure of the foils, which was investigated using the spectral dependence of transmitted light.

I. INTRODUCTION

Solid state strippers used in heavy ion accelerators produce higher average charge states as compared to gas strippers. Therefore solid state strippers are generally used although their lifetimes in heavy ion beams of high current densities are very short.

The development of solid state strippers which means thin self supporting foils followed several courses in the last years. The fact that a higher average charge state can be achieved with elements of lower Z than carbon directed stripper development to the elements Li, Be, B[1,2]. However, the properties of these elements i.e., high affinity to oxygen and humidity, high vapor pressure or high melting point didn't permit any success either in higher charge states or in larger lifetimes of the stripper foils. Tests of these low Z elements as stripper on a thin carbon backing could not satisfy because the stripper lifetime was generally shorter than that of a normal carbon foil. Tests with the carbides of the low Z elements have been without success too[2].

On the other hand no suitable element can be found among those of higher Z-values than carbon either, because its vapor pressure is too high or its melting point too low or the minimum producible thickness of a self-supporting foil is above 20 $\mu g/cm^2$. That is why all recent stripper developments tried to improve carbon stripper foils.

The theoretically longest living carbon stripper foil is expected to be a single crystal of graphite or diamond, because removing an atom from its lattice place takes much more energy than from an intersticial position. The problem is to produce single crystals of carbon with a thickness of $\sim 5\ \mu g/cm^2$.

An epitaxial growth during the vacuum evaporation condensation process is rather improbable as the evaporated carbon atoms are not that mobile on the substrate surface to form larger crystals below substrate temperatures of 20-30% of the absolute melting temperature of carbon (3800 K). A substrate temperature above 600 K, however, results in large mean free paths of diffusion. In this case impurities in the carbon foil, caused mainly by migration of the parting agent molecules into the carbon layer, destroy the better quality of the foils.

Tests were done to raise the energy of the carbon atoms above their thermal evaporation energy before deposition in order to enlarge the mobility along the surface. Such methods for instance carbon arc discharge[3], heavy ion sputtering[4] or chemical deposition[5], might be of great promise if the impurity problem can be solved.

The only way to keep a carbon layer clean at elevated temperatures is to keep it out of contact with any material.

Methods to treat amorphous carbon material - a normal carbon foil is amorphous - in order to get graphite structure are well-known. In the last century first French engineers annealed lampblack to 2800 K and got graphite.

The first results about graphitization process of carbon foils by heat treatment in a high vacuum were reported by

Blue et al(6) 1957 and with more accuracy by Leder et al(7) 1960.

The problem with this Joule effect annealing is the unavoidable heating of the target frame. The carbon film diffuses completely in a stainless steel frame at temperatures above 1200 K.

II. EXPERIMENTAL

1. Preparation and Treatment of the Carbon Foils

We produce our standard carbon foils by evaporation of spectroscopically pure carbon using an electron gun with a magnetic deflection. The carbon is deposited in a vacuum of the 10^{-7} torr range on microscope slides coated with Betaine(8) to float off the carbon films in distilled water. The foils are mounted on stainless steel-or Ta-frames which fit in the standard

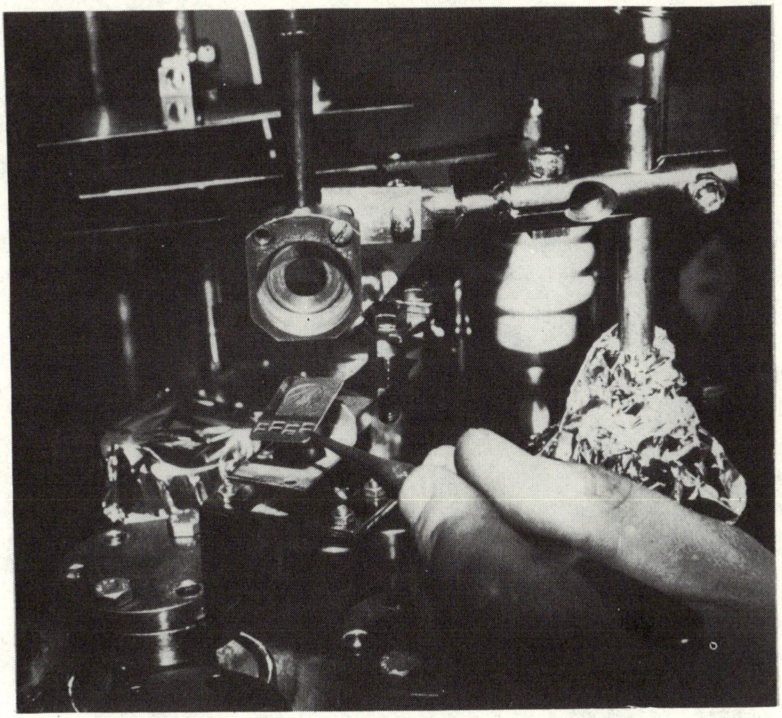

Figure 1. Interior view of the vacuum system to bombard the stripper foils with electrons.

NEC stripper foil changing mechanism. Two of these National Electrostatics Corporation foil changing mechanisms are installed in the terminal of our tandem.

Such carbon foils with thicknesses of $\sim$ 4 μg/cm^2 fabricated by this method have been treated by electron bombardment using two different assemblies. One set up can be seen in Figure 1. A NEC-frame with our carbon foil of 4 μg/cm^2 is positioned on a Ta-aperture with the same hole diameter than the frame. This aperture is mounted on a HV feedthrough and supplied with + 1.5 kV. Under the aperture a tightened Ta-foil filament can be seen which will be heated by low tension. Electrons emitted by this Ta-filament oscillate up and down through the carbon foil until they have deposited all their energy in the carbon foil. The estimated C-foil temperature is not very accurate but it should be close to 2200 K. The bombarding time is only a few seconds, otherwise the frame is heated to temperatures above 1200 K too which must be prevented. Then the carbon target is cooled down by touching with a water cooled finger (we used the head of a quartz crystal oscillator), which is mounted on a push pull feedthrough. The advantage of this electron bombardment compared to one with negative high voltage is an easy set up to get homogeneous heating of

Figure 2. Interior view of the vacuum system with the electrodes to bombard the stripper foils with He^+ - ions.

the whole foil. On the other hand there is no fixed energy. Effects of electron collision cannot be investigated. We also use an electron gun in a Pierce version with negative high voltage for our carbon foil bombardment, but the beam intensity on the foil is not that homogeneous so far.

Another but quite different method of foil treatment is shown in Figure 2. A bombardment with He-ions also showed a prolonged lifetime of our carbon stripper foils, which is so far not understood.

A 1 mm thick stainless steel plate with a rectangular aperture for 5 carbon foils on their NEC-frames is mounted on a HV-feedthrough. Two stainless steel plates are clamped on a rod on ground potential 1 cm under and 1 cm above the plate with the carbon targets of 8 $\mu g/cm^2$ on it. The system is evacuated to the 10^{-7} torr range and then filled with highly purified Helium to 0.4 torr. The plate with the carbon foils is supplied by a negative voltage of 1.5 kV. This results in a positive He current density of about 0.8 mA/cm^2. Twenty-five minutes of bombardment reduce the carbon foil thickness to 4 $\mu g/cm^2$ due to sputtering.

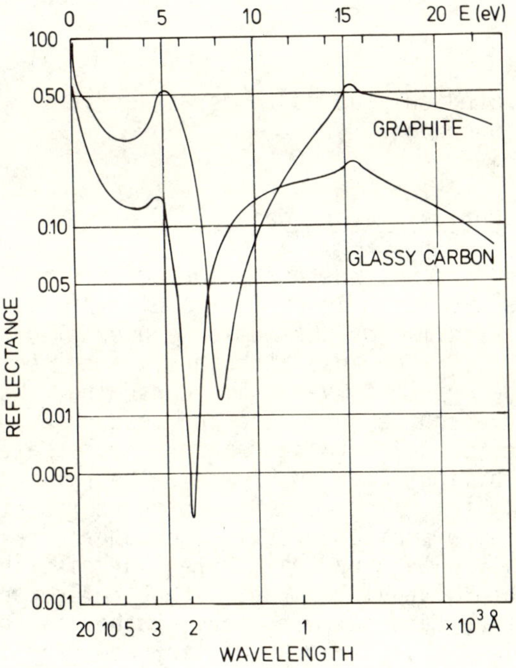

Figure 3. Spectral dependence of the reflectance of graphite and glassy carbon.

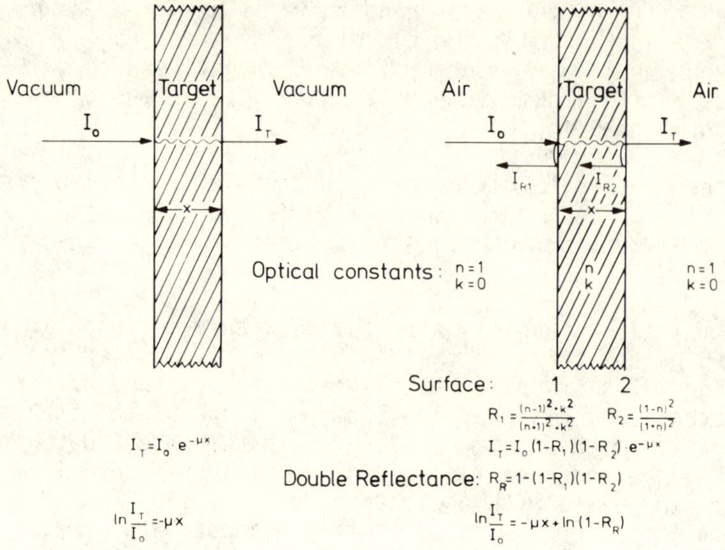

Figure 4. Calculations of optical transmittance and double reflectance.

The foils are heated only up to temperatures in the vicinity of 500 K during this process.

2. Analysis by Optical Techniques

We tried to find out whether a longer lifetime of our special carbon foils was due to a higher degree of graphitization. These tests were not performed by the commonly used electron diffraction pattern method[3,7,9], because its significance seems not to be very high and the method is very time consuming. We used a more empirical method which we suggested and described in 1976[10]. This method bases on reflectance data on carbon materials published by Taft and Philipp[11]. In Figure 3 the reflectivity for monochromatic light is plotted. One graph represents the reflectivity of a natural graphite crystal, the other of glassy carbon a synthetic material with very small unregular distributed crystals. In the visible and ultraviolet regions the reflectance is generally higher for a carbon material with more graphitization. A sharp reflectance minimum moves from 2000 Å for low graphitization to 1550 Å for the graphite single crystal. Reflectivities of thin carbon foils can be calculated using the thickness transmittance relationship deduced earlier[10] and shown again in Figure 4.

R_R the double reflectance is in its trend close to the normal reflectivity. So transmittance data of monochromatic ultraviolet light must show the reflectivity minimum of carbon, the degree of graphitization can be deduced from the transmittance maximum, if the dispersion of the absorption coefficient is low in the wavelength region.

3. Beam Tests

All our standard and special stripper foils have been tested in the terminal of our tandem accelerator. Up to 2.5 μA of S^- - ions generated in a sputter source were injected in the tandem and accelerated by a terminal voltage of up to 10.6 MV. The 90° analyzing magnet selected the 8^+ charge state of the S - ions, which were collected in a Faraday cup. This beam stop was supplied with a positive voltage of 2 kV to measure an accurate current of the analyzed beam without errors due to the loss of secondary electrons. The intensity of the analyzed charge state was measured for each foil until it dropped to 50% of the start value. The lifetime of a standard stripper foil should be shorter in our tandem compared to other tandems, because the ion optics of our accelerator are designed to give a waist of the beam in the terminal in order to compensate the angle straggling in the stripper foil.

III. RESULTS AND DISCUSSION

During the last years many data about prolonged lifetimes of stripper foils have been published. A comparison of the different results is impossible, because there was no unique definition of an absolute lifetime. Some authors published relative lifetimes compared to their standard foils which is useless as long as there is no definition of a standard foil and an easy reproducible procedure to prepare it.

Livingston et al[(12)] published a formula for the reduced lifetime τ of stripper foils,

$$\tau = i \cdot t \cdot A^{-1} \left[\frac{\mu A \cdot \min.}{mm^2}\right]$$

with i the particle current of the incoming beam, t the lifetime of the foil until rupture and A the area of the beam spot on the foil. This formula was used and misinterpreted in many recent publications, which can be explained on an example. A piece of graphite bombarded with a well focused high density heavy ion beam withstands this procedure for an extreme long time, but the

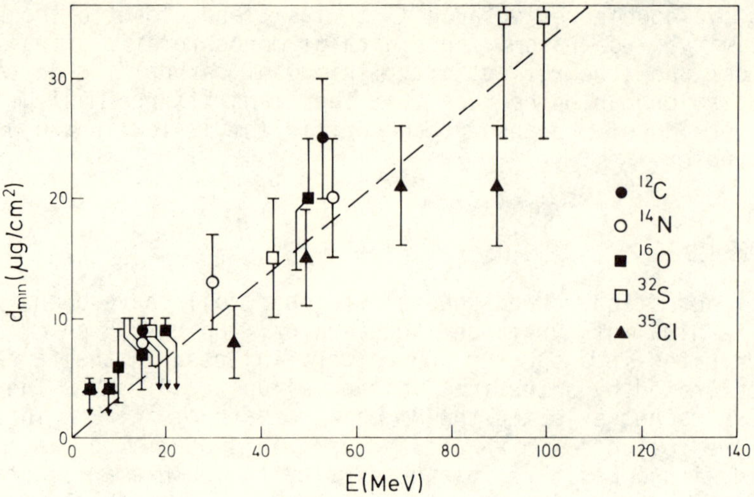

Figure 5. Minimum thickness of carbon foils for equilibrium charge state distribution as a function of ion energy.

interesting stripped beam behind the target is zero. The beam position and spot size during normal accelerator operation are not well-known, they depend on the optics of the system, on the energy and the sort of the ions and last not least on the steering of the individual operator.

The user is interested in the quantity of monoenergetic heavy ions per stripper foil available for his experiment. This means he is interested in a quality of the stripper foil and not in a lifetime. The main quality criterion is given by its thickness: the stripper foil should not be thicker than necessary to achieve the equilibrium charge state distribution, otherwise the quality of the foil is reduced due to low emittance of the beam. Figure 5(13) demonstrates that for normal tandem accelerators with a terminal voltage below 15 MV the foil thickness should be smaller than 5 μg/cm². The intensity of the most probable charge state decreases very slowly for thinner stripper foils, so that the lower straggling in a thinner foil results in higher emittance. For heavy ions up to Ni we measured in our tandem that a 3 μg/cm² stripper foil raises the emittance up to 50% compared to a 5 μg/cm² stripper foil. With this aspect we define a quality factor q in units of Coulomb, which can be measured in every accelerator using the same ions and the identical terminal voltage:

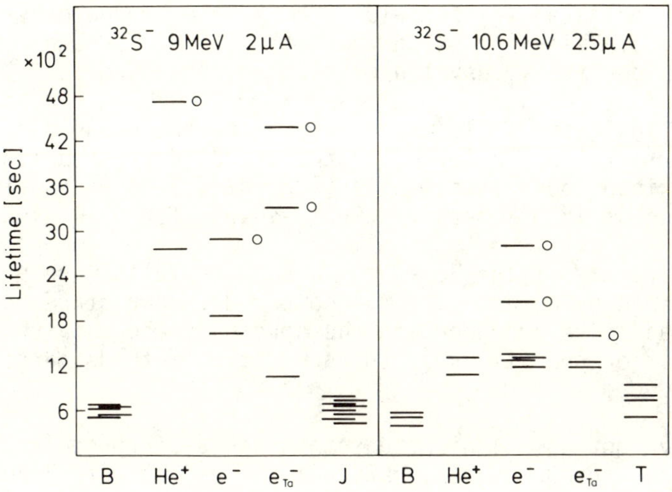

Figure 6. Lifetimes of different treated carbon foils used as terminal strippers for $^{32}S^-$ - beams of 9 and 10.6 MeV and particle currents of 2 μA and 2.5 μA respectively.

$$q = \int_{i_{po}}^{i_{po}/2} i_p \, dt \ [Cb]$$

with i_p the analyzed particle beam current of the most probable charge state and dt the beam time.

The lower integration limit i_{po} means the particle beam current of the most probable charge state at the beginning of exposure. The upper limit of integration is a little bit arbitrary, but normally a stripper foil is not used any longer when the intensity of the analyzed charge state drops below one half of the start value.

Figure 6 shows our first results of prolongation of stripper lifetimes. In order to be able to compare our data with those of other publications the vertical axis is still the lifetime of the foils but the foils indicated with an open circle represent carbon foils, which were not ruptured, but their analyzed 8^+ S-beam reduced to 50% of the start value. All carbon foils were mounted on NEC stripper frames in the terminal of our MP tandem. With B are indicated those foils which were prepared by the standard method described above. The thickness of those foils is generally about 4 μg/cm². The He^+ indicated foils were made by this He-glow discharge method described before. The e^- - and e^-_{Ta} - foils were heated by the electron bombardment procedure. Ta means that the foil is mounted on a tantalum frame.

The J indicated foils (5-10 μg/cm²) were produced by the group of H. J. Maier[14] using the method of S. Takeuchi et al[3]. These foils don't live any longer than our standard foils.

The T indicated foils were prepared by the group of H. J. Maier[14] too. After using the standard preparation technique described above these foils ($\approx$ 5 μg/cm²) were annealed on their frames at 1300 K between two Ta-sheets for 5-10 minutes.

The lifetime of the He^{+}-, e^{-}, and e^{-}_{Ta} - foils were prolonged in the average by a factor of $\sim$ 4 compared to our standard foils, but the statistics are poor and the spread of the data shows that the irradiation conditions or the production methods cannot be well reproduced so far.

We have got some indications that the difference in lifetime for standard and our special carbon foils raises with Z- of the accelerated ion. These results were obtained during our normal beam time from the comparison of the data achieved with O-, S-, Ca-, and Ni beams.

Our optical measurements described above yielded to the

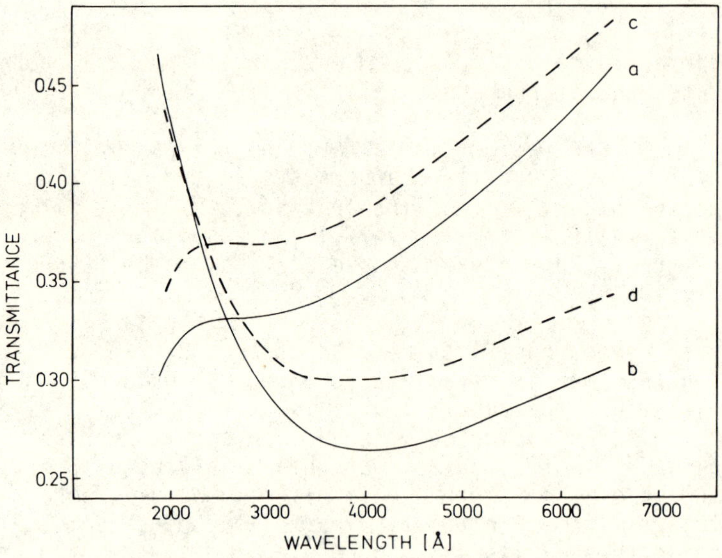

Figure 7. Spectral dependence of the transmittance for a standard 3.7 μg/cm² carbon foil before (a) and after (b) electron bombardment and a standard 3.5 μg/cm² foil bombarded with heavy ions outside (c) and inside (d) the beam spot area.

correlation of graphitization and lifetime of the carbon stripper foils. Figure 7 shows the transmittance of light as a function of wavelength for different treated foils. Curve a) represents a standard 3.7 μg/cm² carbon foil. Curve b) was measured at the same foil after the electron bombardment described above. The transmittance changed drastically in the visible region where it is only about 60% for the electron bombarded foil compared to the standard foil. The reflectivity minimum for wavelengths smaller 2000 Å indicates a high degree of graphitization. Such foils are expected to yield increased lifetimes. Curves c) and d) were measured at a 3.5 μg/cm² standard foil used as terminal stripper, graph c) is at the non-irradiated area of the foil, but graph d) is in the center where a heavy ion beam hitted the foil. The stress lines of the foil indicated, that the irradiation ended shortly before rupture.

The spacing of the curves a) and b) compared to the spacing of c) and d) in the wavelength region of 3000 Å-5000 Å is nearly identical. This demonstrates that the thickness homogeneity was not changed in the irradiated part of the foil (curve d), because graphs a) and b) belong to the identical thickness of one foil. The hypothesis of detectable migration of carbon atoms into the beam spot area on the carbon foil seems to be wrong. The migration theory was created by the visible dark spot on irradiated carbon foils which also occurred when no carbon build up took place. This dark spot distinguishes only the different optical properties of graphite compared to amorphous carbon.

It is not surprising that an electron bombarded carbon foil can give longer stripper lifetimes, because there is no crystal structure difference from the beam spot area to the rim of the foil, no stresses can occur until the crystallites change their size or orientation during the irradiation.

After all these explanations one question is still open, we didn't find any crystal structure in the foils which were treated with He^+ - glow discharge, but these foils also showed a prolongation in lifetime.

REFERENCES

(1) D. Ramsay, Proc. 5th Ann. Conf. of the I. N. T. D. S., LA 6850-C (1976) p. 74.

(2) D. Ramsay, Nucl. Instr. and Meth. 167 (1979) 41.

(3) S. Takeuchi, C. Kobayashi, Y. Satoh, T. Yoshida, E. Tokekoshi, and M. Maruyama, Nucl. Instr. and Meth. 158 (1979) 333.

(4) H. Baumann and L. Wirth, Nucl. Instr. and Meth. 167 (1979) 71.

(5) N. R. S. Tait, D. W. L. Tolfree, B. H. Armitage, and D. S. Whitmell, Nucl. Instr. and Meth. 167 (1979) 21.
(6) M. D. Blue and G. C. Danielson, J. Appl. Phys. 28 (1957) 583.
(7) L. B. Leder and J. A. Suddeth, J. Appl. Phys. 31 (1972) 485.
(8) P. Maier-Komor, Nucl. Instr. and Meth. 102 (1972) 485.
(9) U. Sander and H. H. Bukow, Rad. Eff. 40 (1979) 143.
(10) P. Maier-Komor, Proc. 5th Ann. Conf. of the INTDS, LA 6850-C (1976) p. 150.
(11) E. A. Taft and H. R. Philipp, Phys. Rev. 138 (1965) A197.
(12) A. E. Livingston, H. G. Berry, and G. E. Thomas, Nucl. Instr. and Meth. 148 (1978) 125.
(13) H. Munzer, Jahresbericht 1973, Beschleunigerlaboratorium München (1974) 15.
(14) H. J. Maier, Sektion Physik, Universität München, 8046 Garching, W. Germany.

TRANSMITTANCE VS. WAVELENGTH FOR UNSUPPORTED

CARBON FOILS

John O. Stoner, Jr. and Stanley Bashkin

Physics Department
University of Arizona
Tucson, AZ 85721

We use the optical transmission of carbon foils that we make by arc evaporation as an indicator of their surface density. It's not a perfect method, but it's quick and easy, and reliable at the ten percent level. Other people use it too, sometimes.

We make our measurements at 5461 Å, the green line of mercury. Filters to isolate that wavelength region are not expensive. However, sometimes people need to know the transmission of carbon films at other wavelengths, either because they want to use a different wavelength to monitor thickness, or because they want to

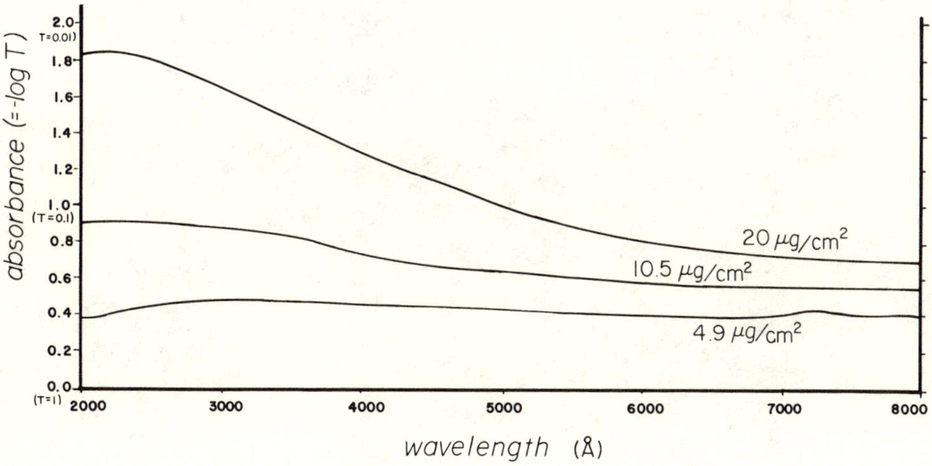

Figure 1.

use the carbon films as optical filters. So we have measured optical transmittances of some representative unsupported carbon foils, and we present them here (Figure 1). There are three things of possible interest on these curves.

First: thin layers serve nearly as neutral filters. The transmittance of a 4.9 microgram/cm^2 layer is nearly constant through the visible, near infrared, and middle ultraviolet.

Second: the slight dip for the 4.9 microgram/cm^2 film near 2000 Å may well be an indication of interference effects.

Third: thicker films are more opaque toward the ultraviolet. If you want a sensitive thickness monitor, it should operate in the ultraviolet region. If you want a monitor that can measure over a wide range of thickness, it should operate in the infrared. We still like to compromise at 5461 Å.

LIFETIMES OF CARBON FOILS DEPOSITED ON ETCHED SUBSTRATES

John O. Stoner, Jr., and Stanley Bashkin

Physics Department
University of Arizona
Tucson, AZ 85721

and

G. Thomas, J. L. Yntema, and P. Den Hartog

Argonne National Laboratory
9700 S. Cass Avenue
Argonne, IL 60439

Many methods have been tried with variable success to make long-lasting foils. Table 1 shows a partial list. If I have left off someone's favorite method, I apologize.

Table 1 Methods of Producing Long-lived (?) Carbon Foils

Slackening
Mounting on oil films
Flashing
Electronic or ion bombardment
Annealing after mounting
Annealing during use
Depositing on heated substrate
Choice of parting agent
Substrate pretreatment
Deposition via cracking in discharges
Deposition via ion plating
Arc evaporation with gas
Arc evaporation with different electrode arrangements

Do all of these methods have something in common? No. But there may be two or three factors, all important in making a foil last a long time. These factors might be:

A. Making a strong, coherent, continuous layer. Thus, good vacuum is desirable for good evaporated foils, and the evaporation rate has to be set to avoid sparks (which make pinholes). The cracking process in hydrocarbon gases is known to make uniform coatings even over rough surfaces.

B. Making a foil slack, loose, or baggy so that it can contract without tearing. Deposited onto an etched surface accomplishes this, whether the etching is done deliberately by reaction of atmospheric water vapor with glass substrates, inadvertently with salt substrates, or by immersion of the substrates into a chemical etchant. The same objective is accomplished by compressing the mount with the foil on it, or by exposing the foil to a brief flash of light, or by heat treatment to permit the foil to relax strains while in use.

C. Making a foil whose molecular structure minimizes shrinkage of the foil under bombardment.

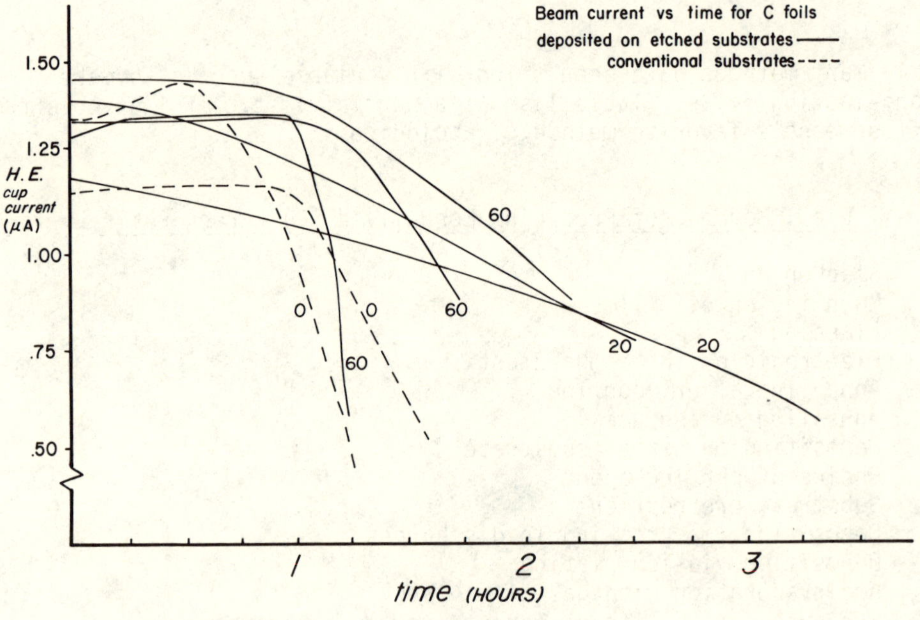

Figure 1.

We suspect that all of the methods listed in Table 1 can be categorized under the above three factors, and that the success that you have in making long-lived carbon foils depends upon how completely you have included these three factors.

We attempted to clarify further the important parameters leading to long foil life by evaporating foils from conventional carbon arcs onto substrates that we know to be rough on a microscopic scale.

We dipped standard microscope slides into a saturated solution of KE 0006, a commercial glass cleaning agent, for 20 or 60 seconds, and rinsed and dried them. We then did our usual soaping and polishing to provide a release layer. We used arc evaporation to produce the carbon coatings.

Figure 1 shows the behavior of foils deposited upon etched substrates compared with foils deposited upon conventional microscope slides. In both cases the parting agent, Creme-Cote, manufactured by James Varley and Sons, St. Louis, Missouri, was used. Beam particle currents of $^{32}S^{-}$ of approximately 0.6 μA at 8.8 MeV were used at the terminal of the Argonne National Laboratory FN tandem accelerator. Beam diameter was about 5 mm and unsupported foil diameter was about 15 mm.

Foils deposited on etched substrates lasted roughly twice as long as conventional foils. This result suggests that the topography of the substrate is important. We suspect that the microscopic roughness of a surface that has been etched, or that is polycrystalline (such as an evaporated-salt surface), gives rise to foils that are slackened on a microscopic scale, and that this is why the surface treatment is important. The great success of the cracking method may then be because it can coat such rough surfaces better than the arc evaporation methods. We are continuing our efforts to find evidence for the relative importance of the three factors that we feel are most important in foil life.

A REVIEW OF DEVELOPMENT WORK ON CARBON STRIPPER FOILS AT DARESBURY AND HARWELL

D. W. L. Tolfree

Science Research Council, Daresbury Laboratory
Daresbury, Warrington WA4 4AD, Cheshire, England

INTRODUCTION

Carbon foils are used extensively as electron strippers for ion beams of MeV energies in tandem accelerators, as exciters at keV energies in beam foil spectroscopy and as backings for nuclear targets. For heavy ions serious limitations exist in foil lifetime.

Since solid carbon foils produce higher mean charge states they are preferable to gases as strippers in heavy ion accelerators. Foils of this type with enhanced lifetimes are very desirable for the new 30 MV tandem accelerator being constructed at Daresbury Laboratory. It was therefore thought worthwhile to carry out development work with the aim of producing long-lived carbon foils and an understanding of foil behavior under heavy ion bombardment. The results of this work have stimulated much interest partly because of the development of new techniques for preparing long-lived foils and also for the comprehensive study that has been made of carbon foils. To place in context, first a brief summary is made of the work done by the Daresbury-Harwell collaboration since 1973. A report of recent and current work is also given.

Early work

In early work foils prepared by the carbon arc technique were used. These were bombarded by 4.8 MeV Ar^{++} ions in a conventional vacuum system (10^{-6}-10^{-7} torr). The foils thickened considerably(1) for beam current densities < 0.1 μAmm^{-2} becoming useless because of excessive scattering of the transmitted beam. For higher current densities lifetime was normally limited by rupture. To overcome the problem of hydrocarbon build-up a clean line vacuum test facility was constructed to operate in the pressure range 10^{-8}-10^{-9} torr[2]. Measurements carried out in this facility confirmed that the limitation was invariably due to rupture.

During irradiation, radial stress lines appear on the foils outside the irradiated area. This is attributed to progressive shrinkage of the bombarded area leading to tightening of the unirradiated area resulting in foil failure. A technique was therefore devised to prepare foils in a slackened condition to lengthen the shrinkage process. Foils prepared by the carbon arc method were slackened by reducing the diameter of the aluminum ring onto which a foil was mounted[3]. The rings were reduced from 17.5 mm to 15 mm diameter by means of a die, giving the foils a very wrinkled appearance. About a factor of ten enhancement in lifetime above that of unslackened foils was obtained. The total contraction of a slackened foil during the irradiation period was such that an increase in the areal density by at least a factor of two was observed in the irradiated area; this was consistent with the shrinkage allowable for a slackened foil. An increased beam scattering and a reduction in the transmitted beam by about the same factor limits the value of foils prepared by this technique.

It was clear that the problem of radiation induced shrinkage could be overcome by the slackening technique. An alternative approach was to make a foil of a material which possessed a lower shrinkage rate. A carbon film with a more ordered structure than the amorphous films normally produced by evaporative methods might be expected to shrink less. Other factors such as the release agent, substrate temperature and deposition process have all been reported to have an affect on foil lifetime[4]. A systematic study was therefore carried out of the influence of many of these factors[5].

Foils prepared by the cracking of ethylene gas in a dc glow discharge had lifetimes on average $\sim$ 20 greater than those produced by the carbon arc process for the purposes of our studies the latter foils were known as 'standard foils'. Large numbers of 'standard foils' were tested in earlier work and their lifetimes and behavior were well known. These types of foils are also extensively used in other laboratories as stripper foils. They show no lifetime dependence on thickness.

The ethylene (hydrocarbon) cracked foils behaved very differently from all other foils under irradiation and were therefore made the subject of further study. In particular they showed a very slow rate of shrinkage and thicker foils (> 15 $\mu g/cm^2$)

Table 1. SUMMARY OF RESULTS FOR ETHYLENE AND STANDARD FOILS

Foil Type	Number Irradiated	Mean Lifetime μA min	Mean Time for Stress Lines to Appear min	Thickness Change	Eventual Cause of Failure
Ethylene, 5-10 μg/cm^2	9	29 ± 8	3	Decrease	Split
Ethylene, 15-30 μg/cm^2	14	71 ± 9	10	Decrease	Hole
Ethylene, Slack, 15-30 μg/cm^2	6	127 ± 20	47	Decrease	Hole
Standard, 3-10 μg/cm^2	17	3.2 ± 0.4	1.7	Increase	Split

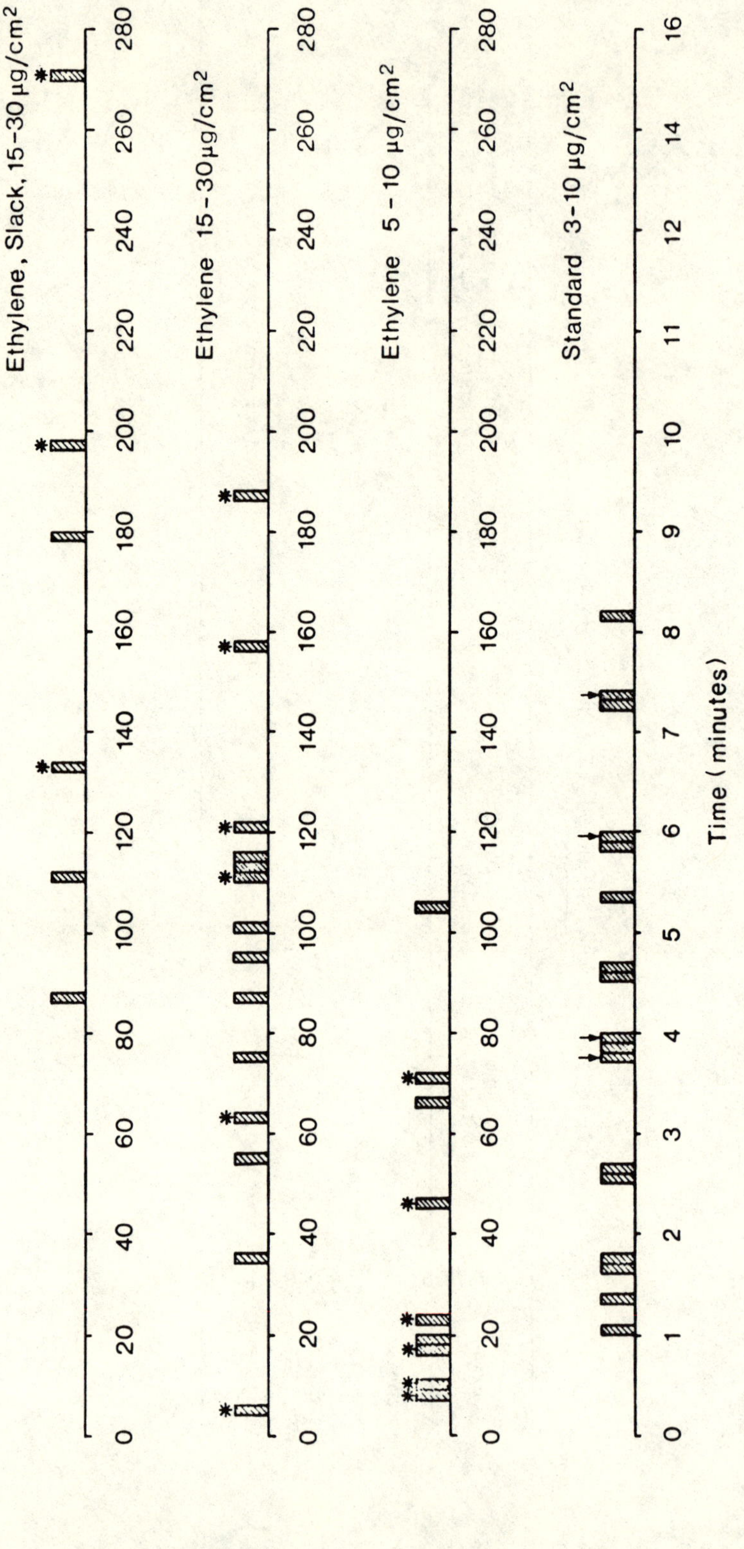

Figure 1. Histograms of lifetimes of standard foils and ethylene cracked foils.

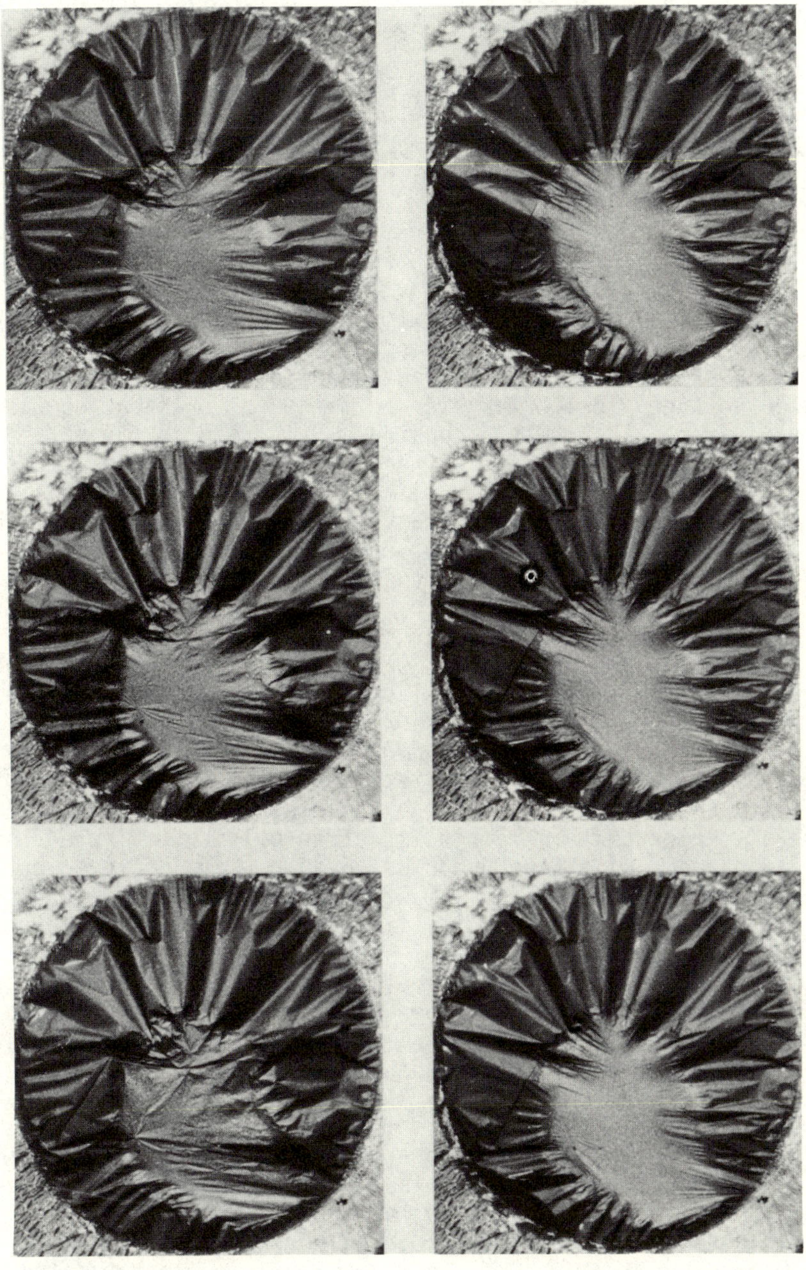

Figure 2. Photographs of a foil prepared from ethylene after different intervals of irradiation.

holed rather than split during irradiation. These foils become thinner during irradiation suggesting that sputtering processes are more than compensating for reduced shrinkage rate.

Work at Daresbury during the last year has been concentrated on ethylene-cracked foils. Further lifetime measurements and irradiation studies have been made. Work is in progress to determine the structure and physical properties of these foils.

Irradiation Measurements

In Table 1 the results of lifetime measurements and observations are shown for the most recent studies of a variety of ethylene-cracked foils. The dependence of lifetime on glow discharge voltage, foil thickness and method of preparation was investigated. There was a considerable spread in the results as shown in Figure 1. Lifetimes were compared as before with those from 'standard foils'. To utilize more effectively accelerator beam time a more damaging 1.2 MeV Ar^+ was used, the current being 0.7 μA in ∿ 3 mm diameter beam spot. Foils were normally 9 mm diameter. Photographs of the foils under irradiation were taken at intervals to give a permanent record of changes; irradiated foil samples were retrieved for subsequent electron beam diffraction studies. The photographs show the progressive shrinkage and developing stress patterns. Slackened ethylene foils in the thickness range 15-30 $\mu g/cm^2$ showed only an increase in lifetime of 1.8 over unslackened foils compared to a factor of 5 for slackened carbon arc foils. Since these thicker ethylene foils fail by puncturing rather than by tearing due to the much slower rate of shrinkage this result is not surprising. Figure 2 shows the appearance of a slackened foil over periods of time up to 270 minutes of irradiation and illustrates the relative weakness of the stress lines. This foil's lifetime was terminated shortly after by the development of holes within the beam spot area. It should be emphasized that our definition of foil lifetime is not necessarily that which would be applicable in the terminal of an accelerator. If foils were to hole rather than split then the beam may still be usable if another part of the foil were used.

The results of the thinner foils suggest a dependence of lifetime on foil thickness (not observed with standard foils). It is not clear at present why this is, maybe the slightly different characteristics are related to the shorter deposition time. The possible effects of gas mixture, pressure and deposition time in the glow discharge process are currently being investigated.

The dependence of foil lifetime on glow discharge voltage and current indicates that foils are improved if the voltage is greater than 3.0 KV. Foils prepared at low voltages < 1.5 KV shatter at the instant of bombardment. This is also the subject of further study since it is helpful in differentiating between 'good' foils and 'bad' foils.

Current Studies

To establish a reliable preparation procedure for ethylene-cracked foils it is vital to have an understanding of the various physical processes involved in the formulation of the film. It is essential to know the structure of the foils and their physical properties to enable important characteristics to be identified.

Electron diffraction studies have been extensively carried out on selected areas of irradiated and unirradiated ethylene-cracked and 'standard foils' as well as 'bad' foils. Although these are still being studied no significant differences have been observed, except for the development of the 002 diffraction ring on some irradiated ethylene foils. Both α-particle scattering techniques and infrared spectroscopy are being used to determine hydrogen-carbon atomic ratios and any hydrogen-carbon bonding respectively. The latter technique has been applied to silicon films produced by decomposing silane gas in a RF glow discharge.

Work carried out by the Japan Atomic Energy Institute and Atomic Energy Ltd.[4,6], Canada on the preparation of evaporated carbon films on heated substrates show that considerable enhancement in lifetime of the foils can result. It is therefore planned to measure the lifetimes of ethylene-cracked foils prepared on heated substrates. Some of these will also be slackened and their behavior under irradiation in the test facility observed.

It has been possible to produce thin ethylene-cracked foils by using an RF glow discharge but was found necessary to back them with thin FORMVAR films to enable successful mounting on foil holders. These will also be tested and compared with standard foils.

The RF technique would enable more uniform large area films to be produced and may give rise to more efficient preparation methods.

Most accelerator applications where heavy ions are to be used require foils of areal density $< 5\ \mu g/cm^2$. Since ethylene-cracked foils are brittle and more difficult to mount than 'standard foils' the efficiency of mounting can be increased to more than 90% by using either FORMVAR or collodion film backings. This is quickly burnt away by the beam during irradiation or can be flashed off by heating prior to installation.

It is intended to measure very thin foils prepared with backings and determine with better statistics values of lifetimes.

Work is also under way to determine the effect of gas mixture and pressure on the efficiency of the cracking process. It is known that foil thickness is related to these parameters but no information on foil structure and lifetime is yet available.

The work on the development of long lived carbon stripper foils reviewed in this paper using the glow discharge technique is a major advance but there are still many factors that need to be studied. Some of these have already been mentioned and are being actively examined. It is not clear at this time if further improvements are possible because the precise atomic structure of the foils and the dependence of lifetime on that structure is not yet known. With many more people now working with these foils and with an increased collaborative effort many of the problems remaining should be solved in the near future.

ACKNOWLEDGEMENTS

Thanks to past and present members of the Daresbury-Harwell collaboration, Dr. N.R.S. Tait, Dr. D.S. Whitmell and Mr. H.H. Armitage and to all the staff of both Laboratories for assistance in various aspects of the work.

REFERENCES

(1) D. S. Whitmell, B. H. Armitage, D. R. Porter, A. T. G. Ferguson, Proc. Int. Conf. on the Technology of Electrostatic Accelerators, Daresbury Laboratory Report DNPL/NSF/R5 1977.

(2) D. S. Whitmell, B. H. Armitage, D. R. Porter, Rev. Phys. Appl. 12, (1977) 1535.

(3) B. H. Armitage, J. D. H. Hughes, D. S. Whitmell, N. R. S. Tait, and D. W. L. Tolfree, Nucl. Instr. and Meth. 155 (1978) 565.

(4) S. Takeuchi, C. Kobayashi, Y. Satoh, T. Yoshida, E. Takekoshi, and M. Maruyama, Nucl. Instr. and Meth. 158 (1979) 333.

(5) N. R. S. Tait, D. W. L. Tolfree, D. S. Whitmell, and B. H. Armitage, Nucl. Instr. and Meth. 163 (1979) 1.

(6) J. L. Gallant, D. Yaraskavitch, N. Burn, A. B. McDonald, and H. R. Andrews, Proc. of the INTDS, Boston 1979.

PREPARATION OF THICK, UNIFORM FOILS FOR RANGE AND RANGE STRAGGLE MEASUREMENTS*

J. M. Anthony

A. W. Wright Nuclear Structure Laboratory
Yale University, New Haven, CT 06520

ABSTRACT

Thick, uniform targets were needed for part of a study of the range and range straggle of energetic (E ≥ 1 MeV/amu) heavy ions. This paper will discuss a method for producing these targets by vacuum evaporation, with a geometry chosen to yield highly uniform targets. Thickness is determined by weighing, and uniformity is measured by comparing the energy loss of alpha particles through different portions of the target. Cu, Ag, and Pb targets have been prepared, with thicknesses up to 9 microns (∼ 10 mg/cm^2) and uniformities ≤ ± 1%.

INTRODUCTION

The range and range straggle of low energy heavy ions can be measured in several ways, for example by ion implanting the projectile in the substrate and doing a ^{4}He backscattering analysis or secondary ion mass spectroscopy (SIMS) measurement on the implanted target. As the projectile energy (and corresponding range) increases, however, the applicability of these methods decreases.

Talk to be presented at World Conference of the International Nuclear Target Development Society October 1-3, 1979.
*Work supported under U. S. Department of Energy Contract No. EY-76-C-02-3074.

Geometrical Uniformity

S (cm)	Z/S	Target Diameter (cm)	% Difference
7.5	1	1.25	0.32
7.5	2	1.25	0.15
7.5	3	1.25	0.07
7.5	1	2.5	1.30
7.5	2	2.5	0.58
7.5	3	2.5	9.28
7.5	1	7.5	11.8
7.5	2	7.5	5.0
7.5	3	7.5	2.5

Figure 1.

A simple way to retain these methods and still measure energetic heavy ion ranges involves the use of thick, uniform targets. Techniques made available by these targets include (1) implanting near the back edge of a uniform target and doing backscattering or SIMS measurements from the rear, (2) measuring the exit energy of the ion beam as a function of incident energy, which allows extrapolation to 0 exit energy and the range, etc. Thus these thick, uniform foils make available several experimental options, including direct access to low energy range measurement techniques.

Method

One procedure available for the production of these foils is vacuum evaporation by resistive heating. The main requirement is a geometry that will produce a constant thickness throughout the target. A single boat at large distances will provide sufficient uniformity, but this uses more evaporation material than is necessary. This mass savings can be quite important for some applications, i.e. preparation of separated isotope targets.

The use of a multiple boat arrangement to improve target uniformity and decrease mass requirements[1] has been used at this lab to produce thick, uniform elemental foils. The geometry chosen involves 4 boats arranged in a square with sides of length S. The source material is placed at the midpoints of the sides of the square, and the target is centered on the square at some distance Z above it. The advantage of the 4-boat system is that nonuniformities due to any boat are compensated for by the others.

Figure 1 shows the results of a simple geometrical calculation based on hemispherical radiation of equal masses located at the midpoints of the sides of the square. The equipment available suggested a square $\sim$ 7.5 cm on a side. Various values of target height (Z/S) and target diameter (assuming circular targets) are given. The fourth column gives the percentage difference between the center of the target (maximum thickness) and the outside edge of the target (minimum thickness). The uniformity of a particular target increases with target height, but the mass required also increases, so the height chosen depends on available mass as well as desired uniformity. The mass savings involved range from 7% (Z/S=3) to 41% (Z/S=1), relatively independent of target diameter. One further factor which enhances greater distances, at least in the case of C backed foils, is that some materials initially boil off large pieces which can easily tear a C backing placed too close to the source.

For the targets discussed here, a height of Z=2S=15 cm was chosen, with a target diameter of 1.25 cm, and evaporation onto 20 $\mu g/cm^2$ C backings. Ideally, the nonuniformity should be $\sim$ 0.15%. However, some materials upon melting will wet the boat containing them, and since the spreading of the material follows no particular pattern, some uniformity is lost upon evaporation. This problem was compensated for by mounting a small electric motor (10 RPM) in vacuum and rotating the target during the evaporation, thus reducing any local nonuniformities.

Heating of all 4 boats simultaneously is accomplished by placing the electrodes on diagonally opposite corners of the square and heating in parallel. This method does not produce the same evaporation rate from each boat. However, if equal amounts (by weight) of material are placed in each boat, and heating is continued until all boats are empty, equal evaporation rates are not necessary.

The foil thicknesses required were 5-10 mg/cm^2, which allowed an accurate thickness determination to be made by weighing the target. The C backings were mounted on aluminum target frames, 2.3 cm square, with a 1.25 cm diameter hole in the center. The frame was then masked, with evaporation followed only on the C covered hole. The target frame was weighed both before and after evaporation, and this information plus the mask area gave the average thickness.

The mask did not allow evaporation onto any part of the C foil directly backed by the aluminum frame. The problem here was that extensive evaporation onto aluminum backed carbon sometimes

Fabricated Targets

Element	Average Thickness (mg/cm^2)	Uniformity (%)
Cu	7.08 ± 0.03	0.3
Cu	4.88 0.02	0.3
Ag	8.39 0.04	0.4
Ag	6.92 0.03	0.4
Ag	4.24 0.02	0.5
Pb	10.25 0.04	1.2
Pb	7.43 0.03	2.5

Figure 2.

caused flaking of the foil, thus making any weight measurement inaccurate. Evaporation only onto unbacked carbon, however, made the targets very fragile, in that the circular foil tended to separate from the target frame after evaporation. Therefore, the frame plus foil was weighed after evaporation, and the foil was then mounted on a second target frame with a 3/8" diameter hole, in preparation for range measurements.

Uniformity was determined by measuring the energy loss of alpha particles through the foil. ^{241}Am was used as a source of 5.5 MeV alpha particles, which were collimated to an area of $\sim$ 2 mm^2 and used to examine various locations on the target. The particles were collected in a Silicon detector, and the energies displayed on a multichannel analyzer.

The energy loss of alpha particles in these thick targets was typically 2-3 MeV, which was large enough to give accurate measurements of the exit energy (ΔE_{out}), and the spread in exit energy (ΔE_{out}) produced by target nonuniformity (as opposed to the width due to energy straggling). The stopping power S(E) of alpha particles in elemental matter is well tabulated[(2)], and thus $\Delta E_{out}/S(E_{out})$ was used to find very accurate values for the thickness variations in these targets. This thickness variation divided by the average thickness gives a value for the target uniformity.

Results

Several elemental targets were made and tested in this fashion, including Cu, Ag, and Pb, as shown in Figure 2. Although none of the targets reached the ideal uniformity of 0.15%, the Cu and Ag targets are quite close. One noticeable effect is the lack of uniformity in the Pb targets. These targets were prepared before the motor was introduced, but the discrepancy appeared too large to be explained away so easily. Several of the targets were examined with an electron microscope in an effort to investigate this. Under 1000 X magnification, the Cu and Ag targets were quite smooth, while the Pb targets showed large crystals, 3-4 microns on a side, scattered randomly across the surface. Since the targets themselves were only 7-8 microns thick, the non-uniformity is explained by these large crystals, and is thus based on the properties of Pb as a target material, rather than the evaporation technique. The Cu and Ag targets did not show this structure, and thus appear to justify the high uniformities measured.

Conclusion

The method of multiple boat evaporation, with each boat containing equal amounts of material, has proved to be quite useful in preparing uniform foils. The uniformities achieved depend to some extent on the target material, but variations less than 0.5% are possible. The techniques involved are quite straightforward, and perhaps the easy production of these targets will encourage the design of experiments that take advantage of their properties.

REFERENCES

(1) Arnison, G. T. J., AWRE Report No. 0-32/67, Atomic Weapons Research Establishment, Berkshire, England, 1967.

(2) Ziegler, J. F., Helium Stopping Powers and Ranges in All Elements, Pergamon Press, New York, 1977.

HISTORY OF TARGET AND SPECIAL SAMPLE PREPARATION AT THE CENTRAL BUREAU FOR NUCLEAR MEASUREMENTS

J. Van Audenhove and J. Pauwels

Joint Research Center of the European Communities
Central Bureau for Nuclear Measurements
Sample Preparation Group
B-2440 Geel (Belgium)

The Central Bureau for Nuclear Measurements (CBNM) was established in 1959 in order to collect data in a limited field of nuclear physics, mainly neutron cross sections. Consequently particle bombardment on targets was an obvious part of the program and the fabrication of targets was considered from the beginning.

In November 1960 the European American Nuclear Data Committee (EANDC) recommended in its meeting at Oak Ridge to establish a major facility for foil and target preparation somewhere in Europe.

In June 1961 on proposal of Dr. J. Spaepen[1] head of CBNM, the European Atomic Energy Commission (Euratom) decided to establish a laboratory which special aim was the fabrication, analysis and distribution of foils, targets and other samples for nuclear measurements, which would also be accessible to outside users. Apart from the preparation of samples by means of known techniques, research was proposed in order to develop and improve existing methods and to develop new ones. From the very beginning the emphasis was laid on well-defined samples requiring more than ordinary skill and equipment for their preparation and assaying. However, the project was somewhat restricted in the sense that no use would be made of remote handling; only alpha-material that could be treated in glove-boxes was accepted. The technical quality of the project was approved by EANDC, and the first part was started immediately (Investment: 360,000 US Dollars, Building costs: 500,000 US Dollars, Recruitment of 24 people).

A number of services were established: chemical and isotopic analysis, alpha- beta- gamma-counting facilities, classical metrology, metallurgy, workshop, radiation control and handling of waste, administration.

1. Preparation Techniques

In 1963 the activity had already developed to such an extent that a number of problems could be tackled. Targets were mainly requested by the national laboratories of France, Belgium, Italy, Germany and the Euratom laboratories but also some demands were received from laboratories outside the European Community.

The preparation techniques used at that time were:

- vacuum deposition using resistance heating for Al, Au, Ag, Cu, Ni, Cr, Co, In, Gd, Pt, NaF, MgF_2.
- vacuum deposition using electron bombardment heating for B, C, Co, Sc, SiO_2.
- electrospraying for U- and Pu targets.
- melting, rolling and machining for pure metals and U/B alloys.
- fabrication of well-defined solutions of boric acid in H_2O and D_2O, salts of Au, Hf, Co, In, Ag, Cs, Th and mixtures of Au and In dissolved in H_2O, Uranium dissolved as nitrate in H_2O.
- powder pressing for S and P samples.

Electrospraying first developed by Carswell and Milsted(2) has been improved at CBNM for the preparation of uranium and plutonium deposits in 1962(3). The materials deposited were mainly U and Pu acetate dissolved in methanol and the deposits prepared varied from nanograms to several milligrams per square centimeter. The substrates used were metal plates or vyns and cellulose (10 to 100 $\mu g/cm^2$) metallized with gold and aluminum (10 to 100 $\mu g/cm^2$).

A limiting factor in the electrospraying of solutions in methanol was the amount of material that could be deposited (a few mg per cm^2). Since 1965 a new modification of electrospraying: "forced electrospraying" has been used for the preparation of very large and thick targets (up to 900 cm^2 surface and 2 mg/cm^2 thickness)(4).

In 1970 it became possible to prepare oxide deposits up to 5 mg/cm^2 by using "suspension spraying"(5).

The improvement of the electrospraying technique can be considered as a significant contribution of CBNM in the target preparation field.

In 1963 electrodeposition of U and Pu were also investigated, but this work was discontinued for the main reason that the electrosprayed deposits were of much better quality.

The high voltage electrophoretic deposition proved to be a versatile coating method for amounts up to 15 mg/cm^2. Since 1964 the method was adapted to our purposes and worked quite successfully for the preparation of various samples of natural and highly enriched uranium e.g., deposits on the inside of graphite tubes[5,6,7].

For the preparation of thicker enriched isotope samples up to several g/cm^2, settling or sedimentation techniques were applied. E.g.: copper-, molybdenum-, rare earth oxide samples (various isotopes), ^{241}Am, ^{241}Pu, ^{237}Np oxide samples. In general these samples were canned in vacuum[8,9].

For the preparation of metallic deposits of rare-earth isotopes - supplied as oxides - thermal reduction in vacuo and the use of different mini-Knudsen cell type evaporation sources have been investigated between 1966 and 1968[10].

Between 1971 and 1974 a considerable amount of ^{252}Cf sources (10^5 to 10^6 f. min^{-1}) prepared by self-transfer have been delivered to European applicants.

For precise measurements at CBNM of the ^{235}U fission cross section, the preparation of UF_4 layers on thin vyns substrates by vacuum deposition has been started in 1970.

For the evaporation in vacuum of reactive metals as Fe, Al, Ti, U and Th, and for the quantitative preparation of pure and homogeneous alloys, the high frequency levitation technique was set up and further developed since 1962[11-17].

A great variety of standard reference alloys, carbides and borides have been prepared since then.

An important reorganization of the CBNM took place in 1973-74. The number of staff involved in nuclear target preparation decreased from about 24 to 9.

In 1976 effort was started in order to increase productivity. By grouping orders from different applicants for a same isotope and with a certain standardization in dimensions, batch production by vacuum deposition of fluorides using multi-substrate holders became sensfull for several actinide isotopes[18,19].

Since then equipment has been set up for "mass" production of ^{239}, ^{240}, ^{242}Pu-, ^{237}Np-, ^{233}, ^{235}, ^{238}U and ^{232}Th-fluoride deposits. Up to 53 high quality deposits with a diameter of about 15 mm or 7 deposits with a diameter of 50 mm can be prepared in one single evaporation run in these facilities.

Another improvement, diminishing final rejections and losses, was the development of polyimide substrate foils (10 to 100 $\mu g/cm^2$) with higher strength, heat, chemical and radiation resistance compared to vyns and formvar[20,21].

The development of the spray-painting technique in 1978 must also be understood in terms of our goal to increase productivity.

In this painting technique an aerograph is used. Compared to classical brush painting rather important thicknesses can be prepared with this technique in one single operation (up to 200 $\mu g/cm^2$). With the use of simple masks, geometrically well-defined deposits, practically without border effects are obtained[22].

2. Assaying Techniques

Accurate nuclear measurements require accurate knowledge of the number of atoms of the nuclide to be measured and of the impurities present in the sample.

The amount of material in these samples (e.g. thin deposits) is in general very small (mg amounts) and yet has to be accurately defined with respect to chemical and isotopic composition.

From the very beginning CBNM payed considerable attention to the accurate assaying of samples. Consequently modern equipped analytical, mass spectrometry and classical metrology laboratories were set up.

As an example of a difficult project realized in 1963 one could mention well-defined and uniform Au, In, Co layers on quartz cylinders (25 mm in diameter, 100 mm long) prepared using a linear arrangement of 5 evaporation sources in order to obtain a uniformity of 3 to 5%. The total mass of the deposits with thicknesses of 0.1 to 0.5 µm was defined with accuracies of 5 to 0.3% respectively[23].

Boron is an important reference substance in neutron measurements. In 1961 CBNM started with the fabrication of standard layers of natural boron by deposition by electron bombardment and weighing in vacuum[24]. The range of boron deposits was 40-150 $\mu g/cm^2$. The maximum accepted thickness variation of the layer was 1% (diameter 38 mm) and the number of ^{10}B atoms per cm^2 had to be known with an accuracy of 0.4%. This high accuracy in ^{10}B mass determination was achieved by careful analytical chemistry, mass spectrometry and by weighing in the vacuum evaporation equipment in order to avoid inaccuracies resulting from handling, oxidation, sorption and desorption effects. Therefore vacuum and even ultra high vacuum balances have been developed at CBNM[25,26].

In the course of precision and accurate measurements of the half-life of ^{234}U between 1969 and 1971 a number of difficulties in analyzing small quantities of uranium were encountered and solved. Quantities ranging from 1.5 to 5 mg were counted by alpha-low geometry and specific activities determined with the results of mass determination by weighing in ultra high vacuum and elemental assay by isotope dilution mass spectrometry. The results agreed within 0.3%[27-30].

For the characterization of evaporated metallic U-layers (about 10 mg) three independent assay methods were used: weighing in ultra high vacuum, constant potential coulometry and isotope dilution mass spectrometry. The results coincided to within 0.3%[30,31,16].

Careful analytical work performed in 1970 showed that LiF layers prepared by vacuum deposition are almost stoichiometric[32].

Between 1977 and 1979 sets of $^{235}UF_4$, $^{239}PuF_3$ and $^{240}PuF_3$ reference fission foils have been prepared by vacuum deposition, defined by relative low geometry alpha counting calibrated by isotope dilution mass spectrometry and verified by different methods in collaboration with NBS. The homogeneity of the deposits and their adherence to the substrate as well as their resistance to different mediums have been investigated[19,33].

In the future similar foils will be made available for the following isotopes ^{233}U, ^{238}U, ^{242}Pu, ^{232}Th and ^{237}Np.

The elements of the rare earth group are of considerable interest in the nuclear field. Isotope dilution mass spectrometry and polarographic methods were used between 1965 and 1970 for the determination of europium. The method has been applied for the accurate determination (1-2%) of 20 ppm and 2 ppm Eu in Al[34].

During the period 1970 to 1971 particular effort went into improving the accuracy of rare earth (RE) determination mainly in small amounts (20 to 50 mg) of Al alloys (1 to 10% RE) needed for neutron dosimetry and in reference solutions.

Between 1964 and 1970 several methods for foil and deposit thickness measurements have been investigated: a pneumatic method suitable for rolled foils as well as beta-absorption, alpha-absorption and soft X-ray absorption for thin films[35].

In 1966 a direct calibration method for a crystal oscillator using a vacuum balance was developed[36].

Between 1970 and 1974 several contributions in the metrology of thin films were made. A technique was developed to evaluate autoradiographs of actinide deposits in conjunction with an isodensitometer in order to obtain a quantitative topography of the deposits[37].

The possibilities and limitations of making non-destructive thickness measurements by the Stylus methods were evaluated when applied to films of soft metals[38,39].

Rutherford backscattering has been used for the determination of film thicknesses[40].

In 1975-1977 a method had been developed to determine the thickness of thin foils by measuring the energy loss of alpha-particles with an energy of 3.17 MeV emitted from a ^{148}Gd source. In vyns foils of 20 $\mu g/cm^2$ thickness variations of ± 0.1 $\mu g/cm^2$ were detectable.

On a set of calibrated Au foils evaporated and weighed in ultra high vacuum (100 to 1500 $\mu g/cm^2$) an uncertainty of about 1 $\mu g/cm^2$ was reached with this method[41].

It is a well-known phenomenon that thick layers prepared by vacuum deposition tend to crack and peel off the substrate. This is associated with mechanical stresses developed during the preparation of the layer. In order to obtain information concerning the influence of various preparative parameters on these mechanical stresses an apparatus was constructed in 1972 and measurements have been performed on Au layers[42].

REFERENCES

(1) J. Spaepen, EANDC (E) 27 "L" June 1961, Installation at the Central Bureau for Nuclear Measurements, Euratom, Geel, Belgium of a Central Laboratory for the Development, Fabrication and Distribution of Foils, Targets and other Samples for Nuclear Measurements.

(2) D. Carswell and J. Milstead, J. of Nucl. Eng. 4 (1957) 5.

(3) K. F. Lauer and V. Verdingh, Preparation by Electrospraying of Thin Uranium, Plutonium and Boron Samples for Neutron Cross Section Measurements in 4π Geometry, Nucl. Instr. and Meth. 6 (1962).

(4) V. Verdingh and K. F. Lauer, Equipment for Electrospraying, Nucl. Instr. and Meth. 49 (1967) 179-180.

(5) V. Verdingh, The Preparation of Layers by Electrospraying and Electrophoresis, Nucl. Instr. and Meth.
(6) V. Verdingh, The Preparation of Samples by Electrochemical Methods, Proc. Seminar on Preparation and Standardization of Isotopic Targets and Foils, AERE-R-5097 (1965).
(7) V. Verdingh and K. F. Lauer, The Preparation of Uranium Layers by High Voltage Electrophoresis, Nucl. Instr. and Meth. 60 (1968) 125.
(8) V. Verdingh, The Preparation of Powder Samples, Nucl. Instr. and Meth. 102 (1972) 431.
(9) G. Del Bino, K. F. Lauer, and V. Verdingh, The Preparation of $^{241}PuO_2$ Samples for Neutron Cross Section Measurements, Nucl. Instr. and Meth. 93 (1971) 205.
(10) H. L. Eschbach and Grillot, Construction et Caractéristiques de Sources Miniatures d'évaporation, Le Vide 24 (1969) 97.
(11) J. Van Audenhove, Vacuum Evaporation of Metals by High-Frequency Levitation Heating, Rev. Sci. Instr. 36 (1965) 383.
(12) J. Van Audenhove and J. Joyeux, Journ. of Nucl. Materials 19 (1966) p. 97-102.
(13) J. Van Audenhove, J. Joyeux, and M. Parengh, Evaporation of Metals and Semi-Conductors in Ultra High Vacuum by Induction Heating, Le Vide, Suppl. No. 136 (1969) p. 69-74.
(14) J. Van Audenhove, Levitation Melting, a Preparation Technique for Reference Samples, Eurospectra 8 (1969) 46.
(15) J. Van Audenhove and J. Joyeux, Sample Preparation by Metallurgical Methods, Nucl. Instr. and Meth. 102 (1972) 409-415.
(16) G. Müschenborn, Die Herstellung Metallischer U-Filme Durch UHV-Aufdampfung für Kernmessungen, Vakuumtechnik 20 (1971) 197.
(17) H. L. Eschbach, Preparation of Standard Layers by Vacuum Evaporation, Nucl. Instr. and Meth. 102 (1972) 469.
(18) J. Van Audenhove, P. De Bievre, and J. Pauwels, Fission Foils and Alloys Containing Fissile Materials Prepared at CBNM 2nd ASTM-Euratom Symposium on Reactor Dosimetry Methods for Fuels, Cladding and Structural Materials, Palo Alto, California, USA, October 3-7, 1977, Proceedings of the 2nd ASTM-Euratom Symposium NUREG/CP-0004, Vol. 3, (1977) 1003.
(19) J. Van Audenhove, F. Peetermans, J. Pauwels, A. Verbruggen, P. De Bievre, and M. Gallet, The Preparation and Characterization of Reference Fission Foils, Nucl. Instr. and Meth. 167 (1979) 61.
(20) J. Pauwels, J. Van Craen, J. Van Gestel, and J. Van Audenhove, Polyimide Substrate Foils for Nuclear Targets, Nucl. Instr. and Meth. 167 (1979) 109.
(21) J. Van Gestel, J. Pauwels, and J. Van Audenhove, Improved Polyimide Foils for Nuclear Targets, World Conference of the International Nuclear Target Development Society, INTDS, Boston, USA - October 1-3, 1979: in press.

(22) J. Pauwels and J. Tjoonk, Spraypainting of Deposits for Nuclear Measurements, World Conference of the International Nuclear Target Development Society, Technische Univ. München, F. R. Germany, September 11-14, 1978, Nucl. Instr. and Meth. 167, 77 (1979).

(23) H. L. Eschbach, Herstellung von Gleichmässigen Aufdamfschichten auf Platten und Zylindern Vacuumtechniek 13, 141 (1964).

(24) J. Van Audenhove, H. L. Eschbach, and H. Moret, Deposition by Electron Bombardment and Weighing under Vacuum of Thin High Purity Boron Layers, Nucl. Instr. and Meth. 24, 465 (1963).

(25) H. Moret and E. Louwerix, Microbalance for Ultra High Vacuum Applications, Vacuum Microbalance Techniques 5, 59 (1966).

(26) H. Moret, E. Louwerix, and E. Sattler, Comments on the Applications and Improvements of a U. H. Vacuum Microbalance, Vacuum Microbalance Techniques 7, 173 (1970).

(27) P. De Bievre, K. F. Lauer, Y. Le Duigou, H. Moret, G. Müschenborn, J. Spaepen, A. Spernol, R. Vaninbroukx, and V. Verdingh, Measurements of the Half-lives of the Uranium Isotopes, Symposium on Neutron Standards and Flux Normalization, Argonne, Ill., USA, October 21, 1970 - AEC Symposium Series No. 23, p. 465 (1970).

(28) P. De Bievre, K. F. Lauer, Y. Le Duigou, H. Moret, G. Muschenborn, J. Spaepen, A. Spernol, R. Vaninbroukx, and V. Verdingh, The Half-life of ^{234}U, International Conference on "Chemical Nuclear Data: Measurements and Applications", British Nuclear Energy Society, Canterbury, GB, September 20-23, 1971, Proceedings: M. L. Hurrell, ed., p. 129 (1971), London Inst. of Civil Engineers, p. 221 (1971).

(29) K. F. Lauer, Assay of Samples for Nuclear Measurements by Counting and Chemical Analysis, Nucl. Instr. and Meth. 102, 589 (1972).

(30) H. L. Eschbach, Preparation of Standard Layers by Vacuum Evaporation, International Symposium on Research Materials for Nuclear Measurements, ORNL-Oak Ridge, Gathinburg, USA-October 5-8, 1971, Nucl. Instr. and Meth. 102, 469 (1972).

(31) Y. Le Duigou, K. F. Lauer, The Coulometric Determination of Uranium: Sampling and Dissolution of 5 to 10 mg Sources Used for Half-lives Measurements, Proceedings of the 14th Conference on Analytical Chemistry in Nuclear Technology, Gathinburg, Tenn., USA, October 13-15, 1970.

(32) Y. Le Duigou, K. F. Lauer, Lithium Content Determination in Lithium Fluoride before and after Vacuum Evaporation, Nucl. Instr. and Meth. 97, 427 (1971).

(33) J. Van Audenhove, P. De Bievre, J. Pauwels, F. Peetermans, M. Gallet, A. Verbruggen, and D. M. Gilliam, Preparation and Characterization of Reference Fission Foils, International Symposium on the Production and Use of RM's, BAM, Berlin, November 13-16, 1979: in press.

(34) P. De Bievre and G. L. Del Bino, Accurate Quantitative Determination of Trace Amounts of Europium in Aluminum, Anal. Chim. Acta 50 (1970) 526.

(35) H. Moret, Thickness Measurements and Mass Determination of Evaporated Layers, Proc. of the Seminar on Preparation and Standardization of Isotopic Targets and Foils, Harwell 1965.

(36) H. L. Eschbach and E. Kruidhof, A Direct Calibration Method for a Crystal Oscillator, Vacuum Microbalance Techniques 5 (1966) 207.

(37) F. Verheyen, W. Dobma, and H. L. Eschbach, Construction and Application of a Simple X Y Scanning Isodensitometer, Journal of Physics E: Scient. Instr. Vol. 4 (1971) 435-437.

(38) H. L. Eschbach and F. Verheyen, Possibilities and Limitations of the Stylus Method for Thin Film Thickness Measurements, Thin Solid Films, 21 (1974) 237-243.

(39) H. Moret and F. Verheyen, Assay of Samples for Nuclear Measurements by Physical Methods, Nucl. Instr. and Meth. 102 (1972) 575.

(40) A. Trapani, A. Berlin, H. L. Eschbach, A. Paulsen, and F. Verheyen, La mesure des épaisseurs des couches trés minces par la méthode de la diffusion de Rutherford, Le Vide, Suppl. 147 (1970) 99.

(41) R. Werz, H. L. Eschbach, and P. Rietveld, Accurate Determination of Energy Loss and Energy Straggling using Alpha Particles, Submitted to: IV Intern. Conf. on Ion Beam Analysis, June 1979, Äarhus, Denmark.

(42) H. L. Eschbach, W. Lycke, and F. Verheyen, Use of Differential Transformer for the Measurement of Mechanical Stresses in Thin Evaporated Films, Vakuumtechniek 22, Heft 8, (1973) 233-238.

SPECIAL NUCLEAR TARGET PREPARATION AT CBNM

J. Van Audenhove

Joint Research Centre
Central Bureau for Nuclear Measurements
(CBNM, Geel, Belgium)

INTRODUCTION

The Central Bureau for Nuclear Measurements prepares samples and targets to meet its own requirements as well as those of other institutions in the European Community mainly.

We report on some more special preparations which were performed last year.

The Preparation of Pd Isotopes Samples

For the measurement at CBNM of the cross sections of fission products, enriched stable Pd-isotopes have been loaned from the US-DOE-ORNL research materials collection in amounts and enrichments as indicated in Table I.

Table I. Weights and Enrichments of Pd-Isotopes

Isotope	Enrichment	Quantities(g)
Pd-104	95.25	20
Pd-105	97.38	13
Pd-106	98.48	46
Pd-108	98.88	41
Pd-110	97.73	18

These Pd-isotopes were available in powder and granular form to be fabricated into 80 mm diameter discs.

Since these isotopes are expensive, the methods used to prepare the targets should generate as little loss as possible and the introduction of impurities should be avoided as well.

We therefore use the following simple procedure.

The base material is first pressed into pellets of about 0.25 cm^3 which are placed onto a copper water cooled cold finger, melted several times by induction heating and furthermore rolled down to plates of 0.2 to 0.37 mm thick and punched into discs.

It turned out that the loss of material remains smaller than 130 mg in all cases. This means between 0.3 and 1%.

The Preparation of ^{208}Pb Samples

To study a number of the possible M1 resonances of ^{208}Pb by means of gamma rays emitted in (p, γ) reactions the University of Groningen needed metallic samples. The Pb enriched at 98.62% in ^{208}Pb was obtained on loan from the US-DOE-ORNL research materials collection in form of $Pb(NO_3)_2$.

This $Pb(NO_3)_2$ is dissolved in water. $Pb(OH)_2$ is obtained by adding an excess of NH_4OH and the yellow Pb_2O_3 is precipitated by adding H_2O_2 drop by drop while stirring the solution.

This Pb_2O_3 filtered on a Buchner filter is first dried at 150°C in air and then transferred to a graphite boat and reduced with H_2 at about 600°C, during 30 minutes. The metal is then allowed to cool to room temperature in an atmosphere of Argon.

The entire batch of 150 g $Pb(NO_3)_2$ has been transformed to metal and remelted and shaped to a plate (40 x 30 x 7 mm^3), a cylinder (Ø : 18 ; H : 8 mm) and a disc (Ø : 21 ; H : 10 mm) with a loss of 0.5%.

$^{240}PuF_3$ Reference Fission Foils

A set of 111 $^{240}PuF_3$ reference fission deposits has been prepared by vacuum evaporation from a resistance heated Ta-crucible. The substrates are stainless steel discs with 0.14 mm thickness and 1.9 cm diameter and a surface of 1.27 cm diameter is covered by $^{240}PuF_3$.

Figure 1. Multi-substrate holder
-Fixed substrate holders in the middle and on the periphery
-Six (6) planetary rotating substrate holders in between

Fifty-three (53) substrates are placed at a distance of about 25 cm from the evaporation source. Three configurations were used during each evaporation run (Figure 1):

-a fixed central substrate holder containing 5 discs.
-Six (6) planetary rotating substrate holders to achieve optimum uniformity. Each holder contains 5 discs.
-Eighteen (18) fixed holders at the border containing 1 substrate each.

A calibrated quartz crystal thickness monitor indicates when the required thickness range is reached.

Series of deposits from 0.6 to 180 $\mu g/cm^2$ have been prepared in this equipment. The deposits are measured by low geometry alpha counting and subsequently some of them were absolutely assayed by destructive isotope dilution mass spectrometry (IDMS). The thus defined calibration factor: alpha counting/IDMS ratio

Table II. Definition of Calibration Factor for ^{240}Pu Deposits

Target N°	Relative α-Counting X_i(cps)	S_i(%)	Calibration µg ^{240}Pu	F=µg ^{240}Pu/cps	Calibration Factor
40L- 5	795.85	0.08	222.17	0.2792	
6	794.38	0.08	221.87	0.2793	
9	608.47	0.09	170.07	0.2795	
10	601.11	0.09	168.30	0.2800	F=0.2798 ± 0.0014
13	121.06	0.09	33.86	0.2797	(0.49%)
16	119.31	0.09	33.04	0.2769	
19	9.965	0.24	2.807	0.2817	
20	10.172	0.24	2.864	0.2816	
22	222.62	0.09	62.45	0.2806	
23	222.62	0.09	62.14	0.2791	

Table III. Amount of ^{240}Pu on Several Foils

Target N°	µg ^{240}Pu Certified	Uncertainty	µg/cm^2	Uncertainty
40L-21	0.757	0.010	0.601	0.008
40L- 7-1	2.551	0.025	2.026	0.021
40L-49	34.47	0.32	27.38	0.27
40L-24	61.85	0.56	49.14	0.48
40L-34	75.59	0.69	60.04	0.59
40L-48	106.53	0.96	84.63	0.83
40L- 1-1	121.51	1.09	96.53	0.94
40L-69	138.54	1.25	110.06	1.07
40L-43	155.91	1.40	123.85	1.20
40L- 9-1	168.88	1.52	134.16	1.30
40L- 2-3	221.53	1.99	175.98	1.71

as mentioned in Table II, allows absolute characterization of the remaining deposits with an accuracy depending on the precision of the low geometry alpha counting for each substrate and of the determined calibration factor and from the accuracy of IDMS (±0.4%).

Furthermore relative alpha counting of 3.14 mm^2 spots of the individual foils indicate variations up to ±3.5% for border targets as shown in Figure 1, up to ±2.0% for central holder substrates and ±0.5% for rotating ones containing 60 to 130 µg ^{240}Pu/cm^2. These variations are expressed as one standard deviation on a single measurement.

Table IV. Isotopic Composition

^{238}Pu:	0.0109 ± 0.0010 at %
^{239}Pu:	0.6724 ± 0.0025
^{240}Pu:	98.4783 ± 0.0088
^{241}Pu:	0.4701 ± 0.0025
^{242}Pu:	0.3677 ± 0.0025
^{244}Pu:	0.0006 ± 0.0003

A border effect of 0.15 mm was observed as well as a slight deviation of the actual deposit diameter (Ø ≈ 12.73 mm instead of Ø ≈ 12.70 mm). The border effect introduces an additional uncertainty of ± 0.37%.

For rotating targets it was experimentally observed that target to target variations in one operation are generally below ± 1.0% for deposits above 2 μg $^{240}Pu/cm^2$.

The amount of ^{240}Pu on each of the 111 foils is given as indicated in the examples in Table III. The isotopic composition of the Pu is given in Table IV.

Quartz Thickness Monitoring

The use of quartz crystal thickness monitoring devices is common practice in vacuum thin film coating systems. The quartz crystal is mounted in a fixed appropriate location in the vacuum chamber and gives rough indication of the obtained mass density on the substrates. Due to possible evaporation anisotropy the ratio between the areal mass density reached on the fixed crystal and that on the rotating substrate, as we use, is not necessarily constant.

We considered it, therefore worthwhile to investigate the possibility of measuring intermittently the mass density deposited on a crystal placed in a rotating substrate holder and as such submitted to the same movements as the substrates on which the coating thickness has to be known. A crystal without leads and cooling has been mounted in place of one of the common rotating substrates during the vacuum deposition as shown in Figure 1 and the areal density of the deposited material has been measured afterwards by plugging the crystal in a sensor head installed outside the vacuum chamber.

The reliability and precision of such a measuring technique has been tried out for gold depositions.

We came to the following conclusions:

a. The variation of the mass density indication due to plugging in and out of the crystal from the sensor head is ±0.4 μg/cm^2

b. After the following operations:

From atmospheric conditions (20 min.) pumping down to 10^{-5} Torr (in 15 min.), returnal to atmospheric conditions (5 min.) and again pumping down to 10^{-5} Torr (in 15 min.), with a water cooling temperature of the crystal of 10 to 15°C the increase in mass density is +0.2 to +1 μg/cm^2. With a water cooling temperature of 24°C this increase is insignificant (≤0.1 μg/cm^2).

c. Condensation of pump oil in vacuum (10^{-5} Torr) could be another source of mass density increase. For a crystal cooled at 10 to 15°C this increase is only about 0.5 μg/cm^2 after 12 hrs. and for a crystal cooled at 24°C no mass density variation can be measured after 12 hrs.

d. Immediately after the deposition is stopped and the shutter closed, the mass density indication increases further, goes through a maximum after about 30 sec. and decreases to a final value being about 2.5 μg/cm^2 more than the value indicated at the end of the deposition.

Summing all variations up we can conclude that it is possible to measure the mass density on rotating substrates to + 3 μg/cm^2.

PREPARATION OF CE-TARGETS

A. H. Bennink and T. W. Tuintjer
Kernfysisch Versneller Instituut
University of Groningen, the Netherlands

ABSTRACT

Metallic samples of separated Cerium isotopes were prepared by chemically converting Cerium dioxide to the sesquioxide form with a high yield and subsequently reduced in vacuo with Thorium metal powder.

The Cerium metal was obtained in a form suitable for preparing 0.1-2.0 mg/cm^2 cyclotron targets.

INTRODUCTION

Westgaard et al[1] have developed a method for the production of rare earth metals by reduction of their oxides with thorium, according to the equation:

$$2RE_2O_3 + 3Th \rightarrow 4RE\uparrow + 3ThO_2$$

This method, however, is not directly applicable to cerium since the inventory form of all cerium isotopes is the dioxide form, CeO_2. Sherman[2] has reported a technique for the reduction of CeO_2 to Ce metal, whereby the CeO_2 was mixed in a hot 6M HCl solution while small quantities of 30% H_2O_2 solution were added. The chemical reaction for this step is given as:

$$6H^+ + 2CeO_2 + 2H_2O_2 \rightarrow Ce_2O_3 + 5H_2O$$

where the Ce^{3+} salt is soluble in the acid solution. Since our aim was to obtain the pure Ce_2O_3 with a high yield, we developed a new method which can be divided in two steps.

Chemical Conversion

The first step involves the dissolution of approximately 100 mg CeO_2 in a 0.24 ml concentrated HNO_3 and 1.5 ml H_2O solution which was kept near its boiling point. The reduction of the CeO_2 was accomplished by carefully adding small quantities (0.1-0.2 ml) of 30% H_2O_2, while the temperature of the solution was maintained between 60-80°C. The chemical reactions for these steps are:

$$2CeO_2(s) + 8HNO_3 \rightleftarrows 2Ce(NO_3)_4 + 4H_2O \qquad (1)$$

and

$$2Ce(NO_3)_4 + H_2O_2 \rightarrow 2Ce(NO_3)_3 + 2HNO_3 + O_2\uparrow \qquad (2)$$

Both reactions take place simultaneously, giving an overall reaction according to the equation:

$$2CeO_2 + 6HNO_3 + H_2O_2 \rightarrow 2Ce(NO_3)_3 + 4H_2O + O_2\uparrow \qquad (3)$$

whereby the $Ce(NO_3)_3$ is soluble and gives a clear colorless solution. The Ce^{3+} ion was then precipitated by adding a solution, consisting of 200 mg oxalic acid in 2 ml distilled water. A 150% excess of the oxalic acid required stochiometrically was used in this step. After warming up the solution to approximately 50°C and subsequently cooling it down to room temperature the precipitate was filtered and thoroughly washed with distilled water, thus giving the finely divided white cerium oxalate powder $Ce_2(C_2O_4)_3 \cdot 9H_2O$ with a 96% yield.

The second step in the chemical procedure was the conversion of the Ce^{3+} oxalate in the sesquioxide form Ce_2O_3. A molybdenum boat (Balzers type 490112) was loaded with the Cerium oxalate and then mounted in the evaporator. After a pressure of 10^{-6} torr was reached the current was switched on and the temperature gradually raised to dehydrate the oxalate. This was done very slowly to prevent the material from spitting out of the crucible and to keep the pressure below 5×10^{-5} torr. When a temperature of about 600°C was reached the crucible was held at the temperature for about 30 minutes and then allowed to cool down to room temperature. The evaporator was brought up to atmospheric pressure and the dehydrated and partly decomposed Cerium oxalate was thoroughly mixed. After evacuating the system again to 10^{-6} torr the high current was switched on and the temperature gradually raised to about 850°C and kept at that temperature for

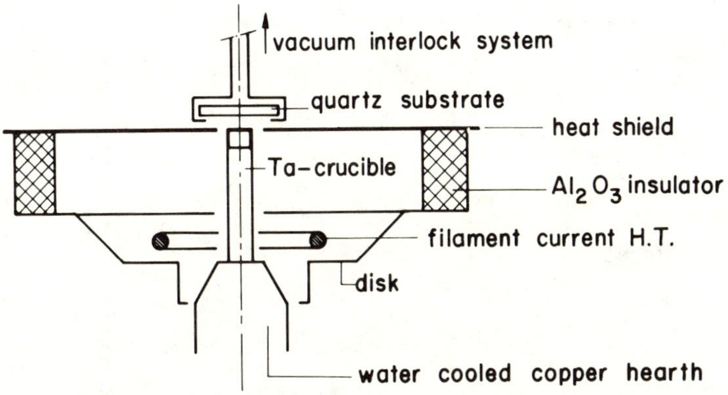

Figure 1. Schematic drawing of the e-gun set up for the reduction and evaporation of Rare Earth elements.

about 1 hour to form the sesquioxide form according to the reaction:

$$Ce_2(C_2O_4)_3 \xrightarrow{850°C} Ce_2O_3 + 3CO + 3CO_2 \qquad (4)$$

The bluish-black Ce_2O_3 powder obtained in this way was allowed to cool down to room temperature. Great care had to be taken not to bring the system up to air while the temperature of the Ce_2O_3 is still too high, since the black Ce_2O_3 turns back to the yellow CeO_2 in air at once above about 150°C. Analyses by coulometric titration were made to establish the completeness of the reaction (4) and revealed a cerium content of 84.8% which indicates a purity of 99.3% of the Ce_2O_3.

Reduction Process

For the reduction of Ce_2O_3 to Ce metal approximately 50 mg of the enriched isotope was thoroughly mixed with about 1.2 times the stochiometric amount of dehydrated thorium metal powder and pressed into a pellet.

The pellet was then placed in a tantalum crucible, 6 mm o.d. and 30 mm high. A conical tantalum stopper with a 2.5 mm hole was placed on top of the crucible and tightened with a slight stroke of a hammer. The tantalum crucible was placed on the watercooled heart of an electron gun[3] and the system evacuated to a pressure of 10^{-6} torr (Figure 1).

After degassing and preheating the high voltage was turned up and the filament current and high voltage were adjusted to give a

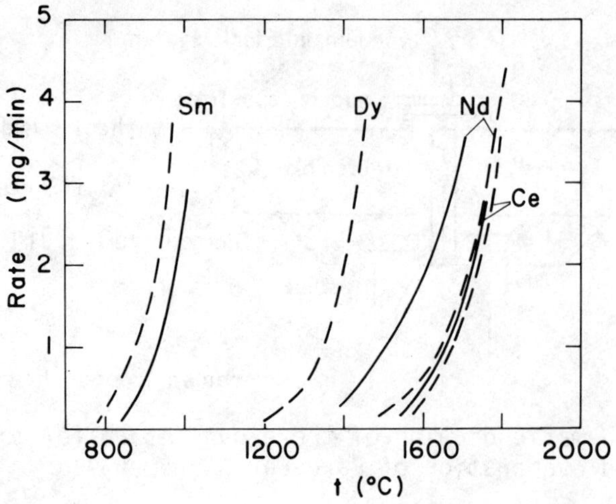

Figure 2. Evaporation rates for some R. E. elements. Dashed lines are the calculated and solid lines the observed rates.

reasonable evaporation rate. An estimate of the evaporation rate was made by using the Knudsen effusion formula[4]:

$$G = \frac{M}{2\ RT} \cdot p \cdot k$$

where G is the mass of evaporant with molecular mass M which escapes through an opening of 1 cm^2 in 1 second, R the gas constant, T the temperature and p the vapor pressure inside the crucible[5].

The Clausing's factor k takes into account the geometry of the opening of the crucible[6]. In our case, with a 2.5 mm diameter opening of 6 mm length, this results in the equation

$$m/t = 9.95 x p x (M/T)^{1/2}$$

where m/t is in mg/min.

The calculated and observed evaporation rates for some rare earth elements are shown in Figure 2.

The temperature of the crucible was controlled by means of a calibrated optical pyrometer and kept at about 1700°C which corresponds to an evaporation rate of about 2 mg/min.

Figure 3. The vacuum interlock system mounted on top of one of the scattering chambers.

Since experiments with the Institute's cyclotron in connection with its Q3D spectrograph do not allow any contaminants such as oxygen and fluorine the substrate chosen was quartz, which was mounted as close to the crucible as possible (1-1.5 mm) (Figure 1). In order to handle the reduced metal under vacuum or a protecting argon atmosphere, the substrate was mounted via a vacuum interlock system (Figure 3).

After the calculated time the high voltage and filament current were switched off and the apparatus allowed to cool down.

The substrate was then transferred via the interlock system into an argon filled glove box.

The collected metal is removed from the substrate by lifting it gently using a surgical scaple. Typical yields of 70-80 percent were routinely recovered in one piece using this method. The thorium contents of typical samples have been determined by X-ray diffraction and are found to be less than 0.8%.

In the case of thick targets (> 1 mg/cm^2), the metal bead was rolled in the argon filled glove box, in the usual manner. For the preparation of thin targets (50-300 μg/cm^2) the framed carbon backings were slid into the target ladder, which was mounted in the interlock system, approximately 10 cm above the crucible. The reduction and evaporation was done simultaneously then. The targets made in this way were stored under vacuum for several months without any sign of deterioration.

REFERENCES

(1) L. Westgaard and S. Bjornholm, Nucl. Instr. and Meth. 42 (1966) 70-80.
(2) J. D. Sherman, Nucl. Instr. and Meth. 135 (1976) 391-392.
(3) T. L. Morgan, private communication.
(4) M. Knudsen, Ann. Phys. 28 (1909) 75.
(5) A. N. Nesmeyanov, Vapor Pressure of the Chemical Elements (1963).
(6) P. Clausing, Ann. Phys. 12 (1932) 961.

PREPARATION AND TESTING OF FERROMAGNETIC Fe, Co,

AND ISOTOPIC Gd FOILS

C. Bichwiller and A. Méens

Centre de Recherches Nucléaires
67037 Strasbourg Cedex
France

ABSTRACT

Techniques for preparing 1 to 2 μm Fe, Co and isotopic Gd foils showing a high degree of ferromagnetism have been developed. Since the magnetic properties of the foils are important for the nuclear experiments in which they were used, an apparatus has been set up which allow prior testing of the foils in the laboratory. In this report, the fabrication process and the testing apparatus will be described.

INTRODUCTION

For the study of hyperfine interactions of excited nuclei recoiling into ferromagnetic media, 1 to 2 μm Fe, Co and isotopic Gd foils have been used as target substrates(1). It is important for the experiments that the foils be highly susceptible to induced ferromagnetism so that full advantage can be taken of the large static hyperfine fields offered by these materials. However making the foils without any precautions will not assure their ferromagnetic properties. Some conditions have to be met, the most important of them dealing with the purity and the structure of the films(2). The techniques used for the preparation of such foils and, importantly to the experimeter, their prior testing in the laboratory will be described in this paper.

Methods of Preparation

For foils of 1 to 2 μm thickness, the usual method of foil

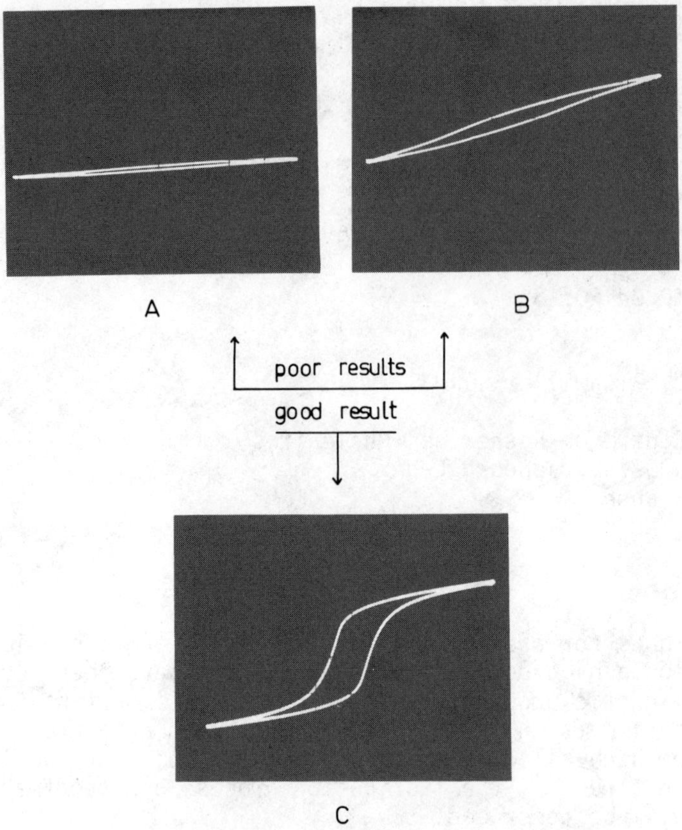

Figure 1. Improvement in ferromagnetic character of foils with annealing after rolling.

Photo A: The hysteresis curve of a 1 μm Gd sample rolled in two directions.

Photo B: The hysteresis curve of a similar Gd target rolled in one direction (improvement over A, but not as good as in C).

Photo C: The same sample as in B after annealing at 450°C for 1 hour.

preparation could be either by rolling or by evaporation. With the attractive rolling method, the foils always must be annealed to have the right magnetic structure. This is necessary even if they are rolled in the same direction as proposed by several workers, as will be seen in Figure 1. For this reason, in our laboratory the evaporation method has been found simpler in practice

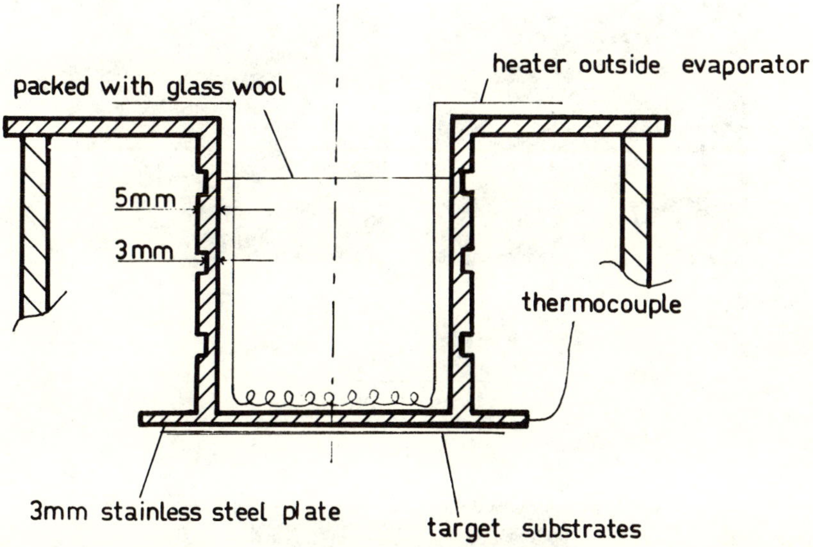

Figure 2. A simple and good system for heating a substrate to 500°C.

for producing natural Fe and Co foils. This is especially so for Co which we found to be very brittle upon rolling. For the isotopic Gd foils we turned to the economical rolling method in spite of the necessity of using the subsequent annealing procedure.

A. Fe and Co

Some conditions have been respected to obtain magnetic foils.

-Concerning the purity, the evaporation has been done in a good vacuum (Diffstack pump, 10^{-6} torr).

-Again concerning the purity and also to obtain the proper crystalline structure, the evaporation speed has to be high. An adequate speed of about 6200 Å/min. (or about 2 min. for 1 μm) was obtained with a 2 kW Varian electron gun. To maintain this speed, a 1 g preformed Fe or Co outgassed bead was used.

-As regards the structure, it was found that the most important point is the substrate temperature. Good results were obtained at 500°C, with the system shown in Figure 2. We found this system useful for reasons related to the fact that there is no heating element and thus no troublesome electrical contacts

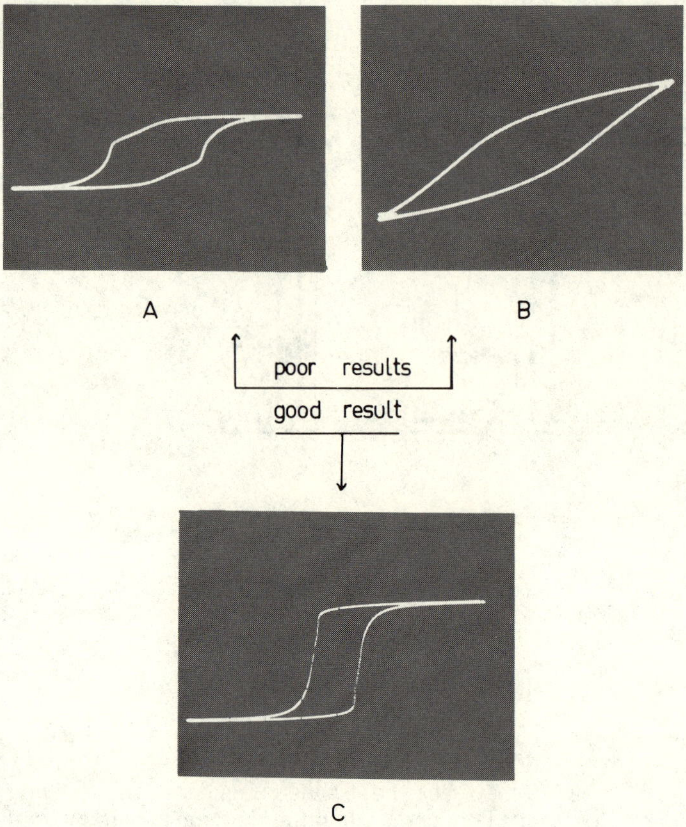

Figure 3. The hysteresis curves of 1 μm Fe foils prepared under different conditions.

Photo A: Evaporation time of 10 min. No substrate temperature modification.

Photo B: Too low a substrate temperature of 400°C. Evaporation time 2 min.

Photo C: Evaporation time 2 min. and substrate temperature 500°C.

inside the bell jar. The cooling has to be slow just as in annealing work.

Figure 3 will show the importance of the conditions. In the case of Fe, the backing for evaporation was 15 μm Cu that was dissolved in Richard's solution. For Co the backing used was 6 μm Al, removed by NaOH.

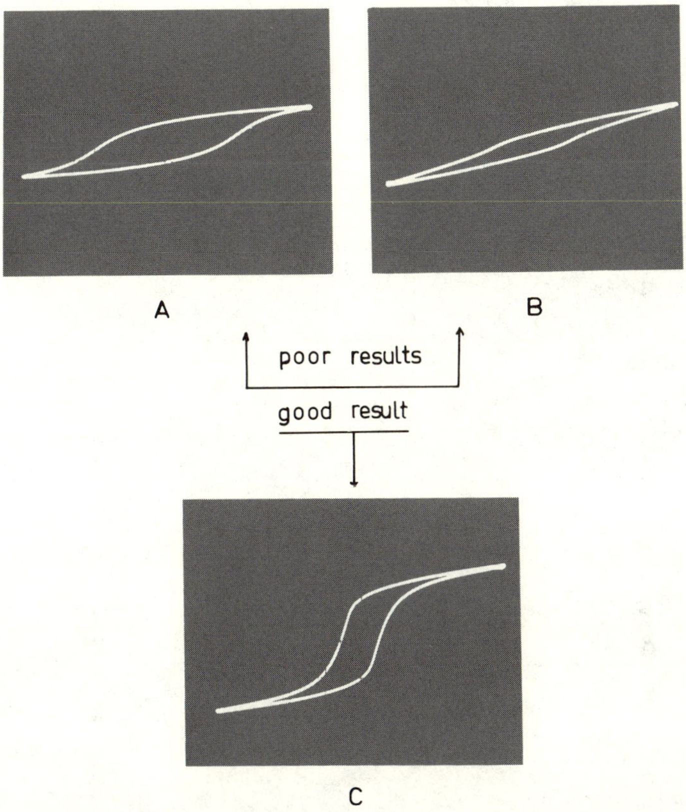

Figure 4. The hysteresis curves of different Gd samples.

Photo A: Curves of a 1 μm Gd foil containing too much Th after distillation.

Photo B: Curve of a 1 μm Gd foil annealing at only 400°C.

Photo C: Curves of a 1 μm Gd foil having the proper magnetic characteristics. Annealing temperature of 450°C.

B. Isotopic Gd

$^{156}Gd_2O_3$ has been reduced and distilled with Th by the method described by Westgaard and Bjornholm (Ref. 3). The product was then rolled with precaution (including elimination of sources of static electricity) because the Gd catches fire very easily. After the rolling, the Gd foil was annealed at 450°C in a quartz

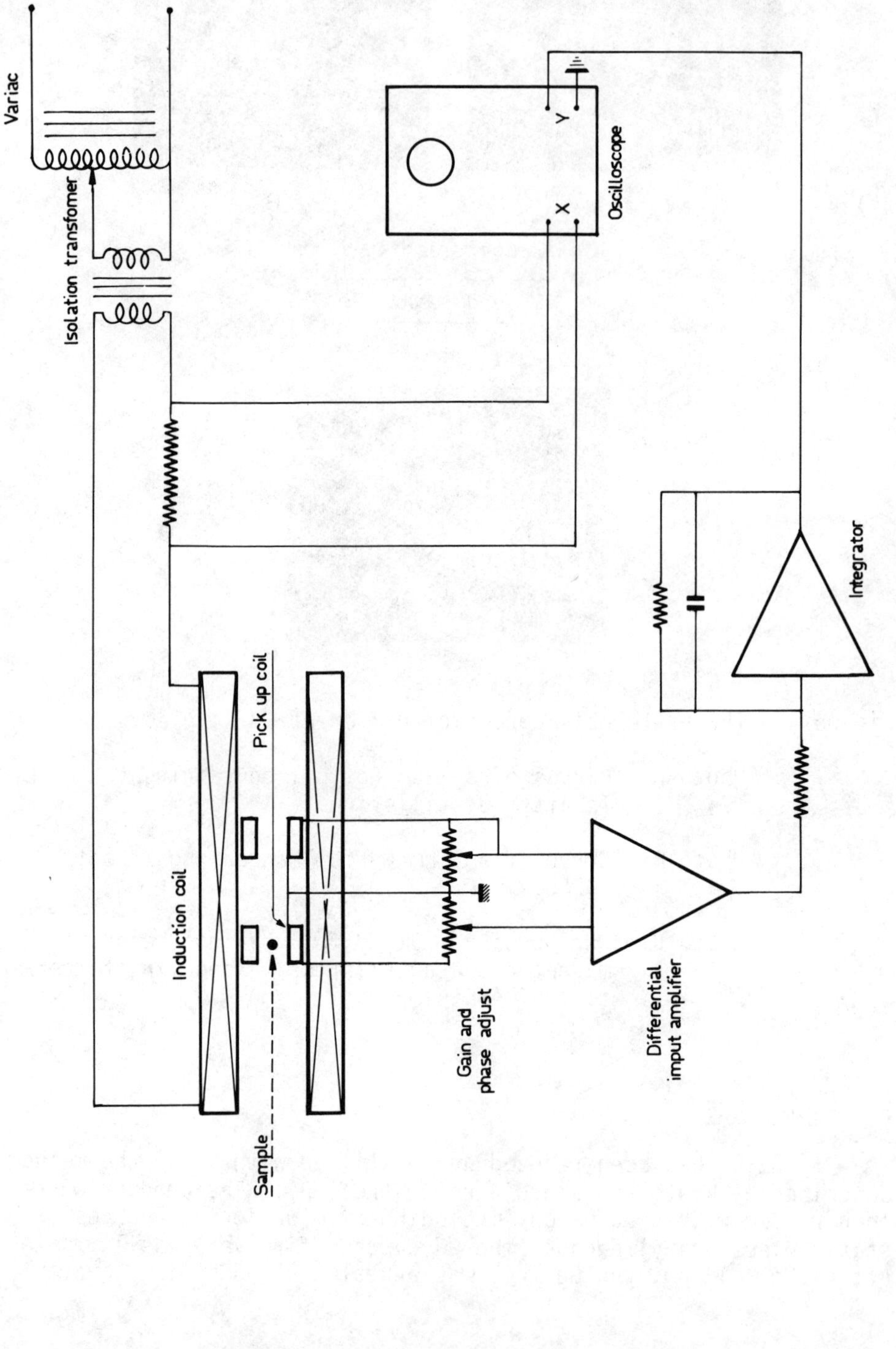

Figure 5. Schematic diagram of the testing apparatus

high vacuum chamber pumped continuously by a turbomolecular pump. Good metallic quality of the end product depends on a good vacuum during the treatment. It is also essential for the magnetic properties that there is no residual Th in the Gd (as was shown by PIXIE analysis) and that the annealing is performed at high enough temperature (Figure 4).

Testing Apparatus

The foils thus prepared were tested in an apparatus which displayed their hysteresis curves. This procedure showed the amount of induced magnetism in a foil placed in an alternating field. This is a measure of the foil's magnetic susceptibility, or in other words, of its ferromagnetic character.

The apparatus is based on an old principle described by Wozek (Ref. 4) after adaptation to small masses by increasing the sensitivity. A schematic diagram is shown in Figure 5. The apparatus is composed of a solenoid through which passes a 50 cycle alternating current. Two smaller pick-up coils are placed inside the solenoid, one receiving the sample and the other serving as a reference. The signal collected by the coils is placed in phase opposition. After amplification and integration, this signal is applied to the Y input of an oscilloscope. At the X oscilloscope input is applied a signal obtained through a series resistance from the solenoid. The time correlation of these two signals, one representing the induced sample magnetism and the other the strength of the inducing field, gives the hysteresis curve of the sample.

Photographs taken of the oscilloscope screen in Figures 1, 3, and 4 show typical results. The desired ferromagnetic quality of the foils is exemplified by "C" of these figures, where it is seen that a small change in inducing field (X-axis) causes a large abrupt and reversible change in induced magnetism (Y-axis). For these susceptibility measurements, the Fe and Co were tested at room temperature, while the Gd was dipped into LN just before making the test to be sure that its temperature remained below the Curie point. All measurements were taken under the same conditions. Foil dimensions were 8 x 6 mm.

The testing apparatus was designed according to calculations. Nevertheless in obtaining best performance, trial and error played an important role. For this reason, we give the following technical specifications, although these values would only be indicative for another apparatus.

Solenoid: 290 mm long; 26 mm i.d.; 1600 turns of 1.2 mm diam. wire; $H = 100$ G.

Pick-up coils: 5000 turns of 7/100 mm diam. wire.

Differential amplifier: Gain = 50 (adjustable); integrator constant = 25 x 10^{-4} (adjustable).

Special Targets

For one experiment a three-layer $^{122}Sn/^{156}Gd/^{nat.}Pb$ sandwich was required which was finally prepared by sticking (Ref. 5). However before sticking the Gd into the 0.1 mm Pb, it was necessary to evaporate 1 mg/cm^2 of Pb on one side of the Gd foil under conditions giving good adherence (glow discharge before and heating at 100°C). The 1.2 mg/cm^2 Sn was stuck on the other side of the Gd after all the Pb had been put on.

The rolling is only needed to stick the Gd to the Pb. Care must be taken not to destroy the magnetic properties of the Gd by an excessive amount of rolling. On the other hand it was necessary to form good mechanical contact between the layers to ensure adequate heat dissipation from the beam spot. This target was cooled by water in contact with the Pb because a temperature rise above 16°C would destroy its ferromagnetism.

In a variation of the experiment, a ^{122}Sn target was mounted as a parallel assembly with a $^{nat.}Fe$ stretched target.

ACKNOWLEDGEMENTS

The authors would like to thank J. Gerber and M. Goldberg for their help and interest in this work and A. Pape for aid in the manuscript preparation.

REFERENCES

(1) C. Broude, M. B. Goldberg, A. Zemel, J. Gerber, J. Kumbartzki, and K. H. Speidel (to be published).

(2) M. S. Cohen, Handbook of Thin Film Technology, Ed. by L. I. Maissel, and R. Glang, McGraw-Hill, New York, 1970, Chapter 17.

(3) L. Westgaard and S. Bjornholm, Nucl. Instr. and Meth. 42 (1966) 77.

(4) J. Wozek, Thèse de docteur d'Université, Université de Paris Sud, Orsay, 1976 (unpublished).

(5) L. Sapir, Proc. of the Conf. of the Nucl. Target Development Soc., Universität und Technische Universität München, 1978.

THICK NOBLE GAS TARGETS PREPARED BY ION IMPLANTATION

W. Cole and G. W. Grime

University of Oxford
Department of Nuclear Physics
Keble Road, Oxford, U. K.

ABSTRACT

Targets of up to 5 μgcm^{-2} He and 15 μgcm^{-2} Ne have been prepared by ion implantation into Ta. In order to achieve the large doses, the targets were cooled to 77°K during the implantation and the beam was swept over the target to avoid hot spots. The energy used was 60 keV at currents in the region of 6 μAcm^{-2}.

The technique of preparing nuclear bombardment targets by ion implantation is not new[1,2], but has not been widely used because the amount of implanted material which may be collected is small compared with conventional target preparation techniques. In the case of inert gases, however, implantation is the only method available for preparing foil type targets, and so it would be desirable to try to increase the amounts of these materials which may be collected.

This paper is a preliminary report of the work in progress at this laboratory to prepare targets of inert gases, particularly He and Ne, containing useful concentrations of target material.

Background

The amount of material which may be introduced into a host lattice by implantation is limited by two processes; self-sputtering and diffusion. In the case of self-sputtering, a concentration of implanted material is reached where each impinging ion sputters an atom of the material already implanted, thus there is no net increase of concentration. Diffusion is the process whereby the

Table 1. Saturated Values of Ne Implanted at 45 keV Into Foils of Various Elements [Data taken from (3)].

Element	Saturation Value (μgcm^{-2} of Ne at 45 keV)
Mo	4.7
Ta	5.0
W	4.5
Cu	1.7
Ag	0.8
Pt	1.7
Au	1.0

implanted atoms may migrate to the surface and leave the foil. This process is temperature dependant and takes place continuously.

These effects are interdependant to some extent, since if the energy of the impinging ions is increased, the self-sputtering will increase, but since the implanted ions will penetrate further into the lattice, the effects of diffusion will be reduced.

The maximum implanted concentration which may be achieved under a given set of conditions is called the saturation value, and has been measured by Almen and Bruce[3] for various combination of target foils. Table 1, taken from reference[3] shows the saturation values of Ne implanted at 45 keV into various materials which may be considered as suitable backings.

It will be seen that the refractory metals are capable of holding significantly more neon than the more ductile metals, and so these materials should be used where possible.

Almen and Bruce also show that the saturation value increases with the energy of the impinging ion, thus the maximum beam energy should be used.

Since diffusion is temperature dependant, it is possible to reduce its effects by cooling the foil. The available data[4,5] indicates that the diffusion constants of gases in metals are very strongly temperature dependant, with a variation of the form[4]

$$\log D = C - \frac{B}{T}$$

where D is the diffusion constant

T is the temperature (^{o}K)

and B and C are constants

The values of B are in the region of 10^3 to 10^4 (e.g. for H_2 in Mo, B=4417). Thus if we take two extremes, the temperature of a red-hot foil under intense ion bombardment ($\sim 800^{o}K$) and the temperature of a foil cooled by liquid nitrogen ($\sim 100^{o}K$), we find that the diffusion constant changes by a factor of approximately 10^{16}! Thus cooling the foil is very important.

To summarize, in order to collect the maximum possible amount of material, the following points must be observed:

1) The host material must be carefully selected
2) The maximum possible ion energy must be used
3) The foil must be cooled to a low temperature.

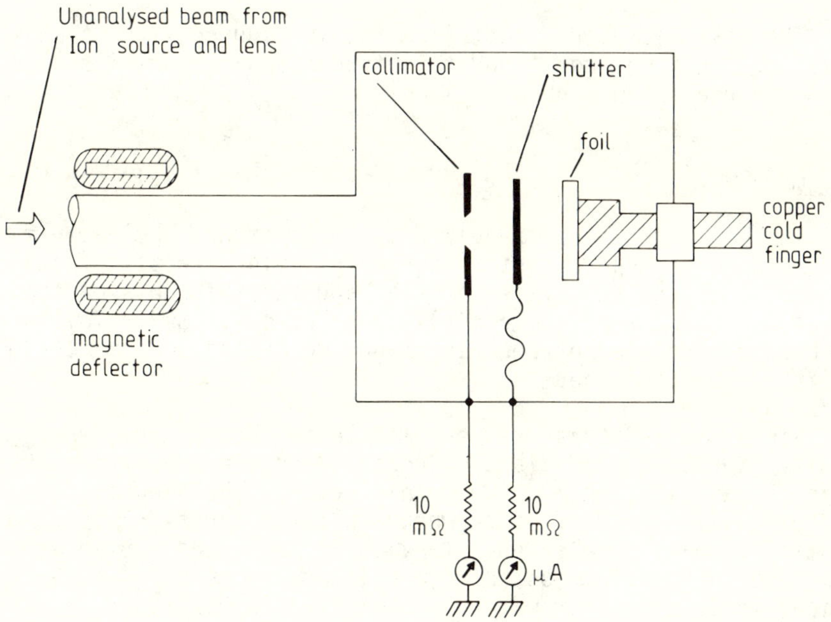

Figure 1. Schematic diagram of the apparatus used to prepare implanted targets.

Apparatus

The apparatus used is shown schematically in Figure 1. The ion source used is a Harwell sputter source(6), which is used as an arc source for gases. In principle, however, any simple ion source could be used, since only a few microamps are required. The beam passes through a simple electrostatic lens which forms a diffuse spot of a few cm diameter in the target chamber. The maximum energy which may conveniently be used is 60 keV.

In order to achieve the cooling of the target, the foil is mounted on a stub on the end of a copper cold finger which is immersed in liquid nitrogen. In order to avoid local heating of the foil, the beam intensity is kept low ($\sim$ 6 μAcm^{-2}), and as a further precaution, the beam is rastered over the foil by means of the magnetic deflection coils. This reduces the effect of any non-uniformity of the current distribution which could give "hot spots" on the foil.

The target area is defined by a 10 mm dia. collimator and a shutter is used to monitor the current periodically during the implantation. The beam current is measured through a large resistor ($\sim$ 10MΩ) to help reduce the effect of secondary electrons.

The beam was used direct from the ion source without mass analysis. This was not found to introduce any impurities as long as the source gas was pure.

Helium

The He targets were required to be stretched optically flat using a plunger type target holder. This places restrictions on the choice of the substrate material and the thickness of the foil and it was found that commercially available 1.5 μm Ta foil offered the best compromise between resistance to radiation damage and the ability to be stretched.

Since the He atoms are very mobile in the lattice, they readily migrate towards lattice defects and eventually form high pressure gas blisters(7). When this occurs, the amount of He in the foil falls rapidly, since when a blister bursts, a large quantity of gas is released. In view of this, the implantation is terminated at a point just before blistering occurs. For 60 keV He ions in Ta at approximately 5 μAcm^{-2}, this was found to take 10 to 12 hours.

The amount of He in the foils was investigated by Rutherford backscattering (RBS). The foil is bombarded with a beam of 1.5 MeV protons and the energy of the backscattered particles is measured

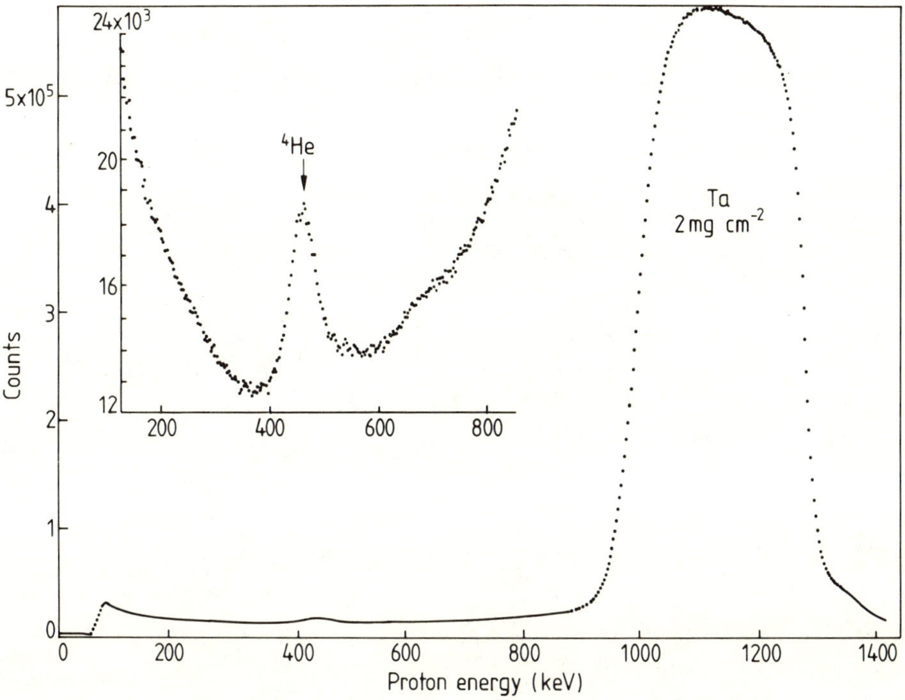

Figure 2. Rutherford backscattering spectrum of He in Ta target.

with a surface barrier detector. Since the energy loss of the backscattered protons is greater for light target atoms, the spectrum of backscattered particles gives information on the constitution of the target. A typical spectrum is shown in Figure 2. A large, broad peak represents scattering with very little energy loss from the Ta atoms, while a small peak represents scattering with large energy loss from the He. From a knowledge of the RBS cross-sections and the number of counts under the He peak, it is possible to estimate the total mass of He in the foil. It can also be seen from Figure 2 that there are no obvious impurity peaks.

Initial results with the foil water cooled showed a maximum implanted dose in the region of 2.5 to 3 μgcm^{-2}. With the foil cooled by liquid nitrogen, the maximum dose has increased to 5.5 μgcm^{-2}. In order to prolong the life to the target, it is necessary to cool it during use, so that the heating by the experimental beam does not drive off the implanted gas.

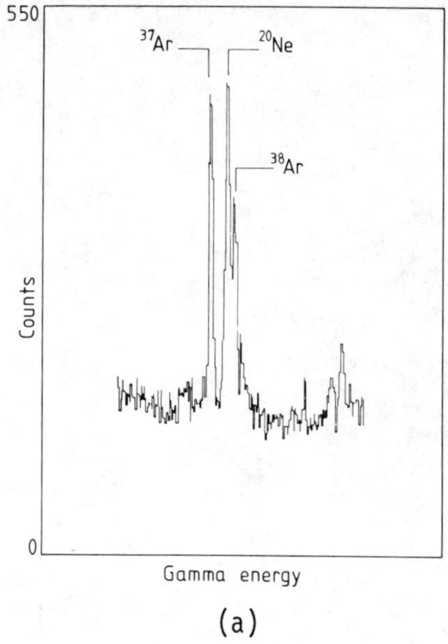

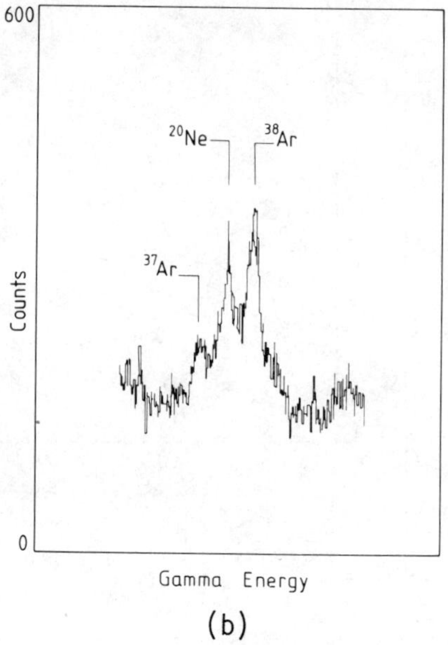

Figure 3. Gamma spectra of implanted target and ^{20}Ne gas target bombarded with 35 MeV ^{20}Ne.

a) Implanted target;
b) Gas target.

Neon

The requirements for the Ne targets allowed the use of more substantial foils and 10 μm Ta was used. The conditions of the implantation were identical to those for He except that the foil saturated after a shorter period (6 to 7 hours). The neon targets could not be measured by RBS, since the energy loss in the thick foil causes a broad Ta peak which overlaps the Ne peak. However, it has been possible to compare the nuclear reaction using an implanted target with that from the same reaction using the Oxford gas target(8). This comparison is shown in Figures 3 a) and 3 b), which show the spectrum of gamma rays emitted from implanted target and the gas target under bombardment by 35 MeV ^{20}Ne ions. The central peak in each triplet is a gamma ray emitted from coulomb excited ^{20}Ne, which could therefore be emitted either from nuclei in the target or those in the beam. The outer peaks, however, are gamma rays emitted from the products of reactions between the Ne in the beam and that in the target. Thus the ratio between the central and outer peaks is a rough indication of the amount of Ne in the target. It will be seen that the peaks in both cases are of similar intensity, thus we may imply that the implanted target contains a similar quantity of Ne to the gas target, which in this case was $\sim$ 15 μgcm^{-2}. This is only a crude comparison, since the geometry of the experiments were not identical, but it gives an indication that usable targets may be produced by implantation. Also evident from Figure 3 is the increased width of the peaks from the gas target. This is due to Doppler shift caused by the motion of the target atoms, which is much greater in the gas target.

Conclusion

Preliminary results have shown that useful targets of He and Ne may be prepared by ion implantation using simple apparatus. It has been found that by preparing the targets at low temperatures, the amount of collected material may be increased.

ACKNOWLEDGEMENTS

The authors wish to thank Mr. J. Billowes, who is using the He targets in nuclear lifetime measurements. He obtained the RBS spectra and made many helpful comments. They also acknowledge the cooperation of Dr. N. Poffé, who kindly supplied the spectra of Figure 3.

REEERENCES

(1) Santry, D. C., Proc. INTDS Target Conference (Chalk River, 1974), p1.
(2) Arnell, S. E., Nucl. Phys. 24, 500 (1961).
(3) Almén and Bruce, Nucl. Inst. and Meth. 11, 257 (1961).
(4) Dusham, S., Scientific Foundations of Vacuum Technique, 2nd ed., p. 570 (Wiley, New York, 1962).
(5) Kohl, W. H., Handbook of Materials and Techniques for Vacuum Devices, p. 612 (Reinhold, New York, 1967).
(6) Hill, K. J. and Nelson, R. S., Nucl. Instr. and Meth. 38, 15 (1965).
(7) Terreault, B., St. Jacques, R. G., Veilleux, G., L'Ecuyer, L., Brassard, C., and Cardinal, C., Helium Irradiations of Copper at 1 to 25 keV: Range Profils, Reemission, and Blistering.
(8) Allen, K. W., Dolan, S. P., Holmes, A. R., Symons, T. J. M., Watt, F., Zimmerman, C. M., Litherland, A. E., and Sandorfi, A., Cryo-pumped Gas Target for the Study of Radioactive capture reactions.

IMPROVED POLYIMIDE FOILS FOR NUCLEAR TARGETS

J. Van Gestel, J. Pauwels, and J. Van Audenhove

Joint Research Centre, Central Bureau for
Nuclear Measurements
(CBNM, Geel, Belgium)

I. INTRODUCTION

At the last Conference of the International Nuclear Target Development Society we reported on a preparation method of thin polyimide foils for nuclear target fabrication via amide-acid spreading and in-situ polycondensation on NaCl coated glass plates(1).

The foils obtained showed excellent chemical-, heat- and radiation resistance properties as well as acceptable mechanical strength in comparison with other foils of equivalent thickness. Heat resistance tests however suggested that improved mechanical properties could probably still be achieved by applying a more appropriate thermal treatment to the precondensated foils.

II. OPTIMIZATION OF THE POLYCONDENSATION PROCESS

A series of pressure tests were carried out as described in ref. (1) on approximately 40 $\mu g/cm^2$ polyimide foils prepared under different conditions. The comparison of their mechanical strength is however complicated by the fact that it is difficult to prepare foils of accurately predefined thickness, and that even starting from the same amide-acid solution the applied heat treatment may influence the foil thickness. To avoid interpretation problems and decrease the number of experiments to be carried out, foils with thicknesses between 35 and 50 $\mu g/cm^2$ were taken into consideration, but their results were normalized using the rupture constant as defined by Chen(2):

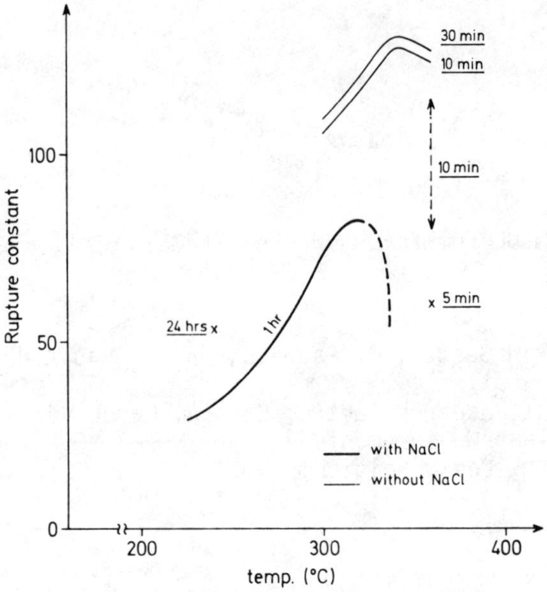

Figure 1. Influence of baking conditions on foil strength (pressure test applied to foils on rings of 20 mm internal diameter).

$$R = \frac{p \times D}{t}$$

R = rupture constant
p = pressure in cm Hg
D = self-supported foil diameter (cm)
t = thickness in $mg\text{-}cm^{-2}$

The rupture constants of polyimide test foils baked during different times at different temperatures, with and without NaCl as release agent, are summarized in Figure 1. From this figure it appears that the foil strength increases with rising baking temperature and, to a smaller extent, with baking time. However, when NaCl is used as a release agent, a temperature increase above 300°C makes the release of the foils progressively more difficult and this has the effect of causing a reduction in the foil strength. The use of shorter baking times can improve this situation, but e.g. with a 10 minute baking at 360°C very irregular results are obtained both for the mechanical strength of the foils and their production yield which may vary from 0 to 100%.
These problems are significantly reduced when no release agent is used. In this case excellent foils are obtained, with a yield of practically 100%, by baking them at 340°C during 10 to 30 minutes.

III. MODIFIED FOIL PREPARATION METHOD

Based on the above experiences, a modified polyimide foil preparation method has been adopted.

1. Cleaning of the Glass Plates:

A glass plate minimal rugosity is degreased and thoroughly rinsed with tap water, distilled water and methanol. It is important to eliminate all dust particles during this operation. Before use, the glass plate is therefore put on a centrifuge where it is further rinsed with methanol and dried by centrifuging.

2. Preparation of the Amide-Acid Solution:

Stoichiometric quantities of 1, 2, 3, 4 - benzenetetracarboxylicdianhydrid and 4,4' - diaminodiphenylether are dissolved in a weighed quantity of NN' - dimethylformamide. The concentration of this solution is chosen to correspond to the desired foil thickness(1) and stirred magnetically during 4 hours, after which the concentration is corrected with some drops of NN' - dimethylformamide to compensate for solvent evaporation.

The solution is then filtered and stored for the night in a well-closed vessel.

3. Foil Preparation:

Approximately 5 ml of the amide-acid solution are deposited in the middle of a clean glass plate and spread over it by centrifuging for one minute at 6500 rpm. The glass plate is then heated for 4 hours at 100°C to enable complete solvent removal and baked for 15 to 20 minutes at 340°C.

After cooling, the foil is floated off in a distilled water bath and picked up on a target ring.

IV. RESULTS

1. Production Yield

Polyimide foils have been prepared as described above on glass plates of 10x10 cm². They were cut in 9 squares and picked up on aluminum rings of $\emptyset_{ext}$ = 40 mm, $\emptyset_{int}$ = 20 mm, th = 1 mm. When 9 foils could be obtained from one glass plate the yield was

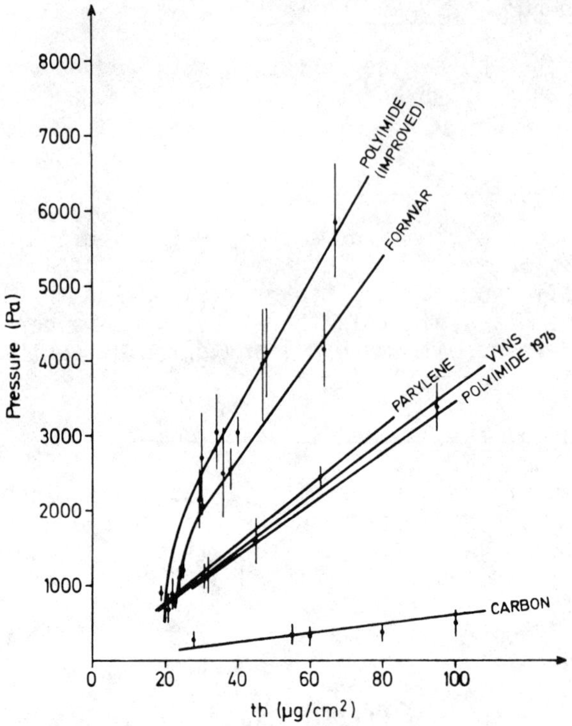

Figure 2. Results of pressure tests on foils mounted on Al-rings of 20 mm internal diameter.

considered to be 100%. This was generally the case above 30 μg/cm^2. Below 20 μg/cm^2, however, the yield dropped practically to 0%, because the foils could no longer be released from the glass plate.

These experiments furthermore confirmed that it is difficult to predict the real foil thickness to better than ± 10 μg/cm^2.

2. Mechanical Strength

Compared to the foils described in ref. (1) a significant improvement in mechanical strength has been obtained. This is

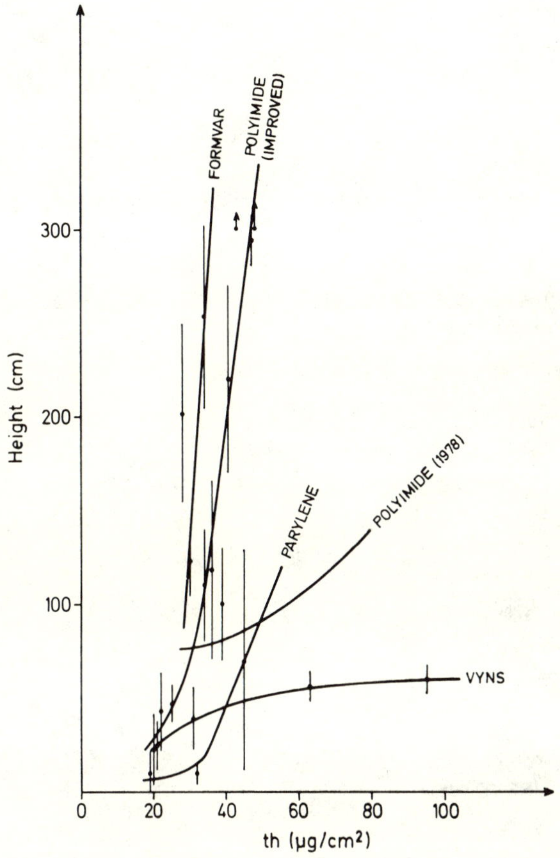

Figure 3. Results of dropping tests on foils mounted on Al-rings of 20 mm internal diameter.

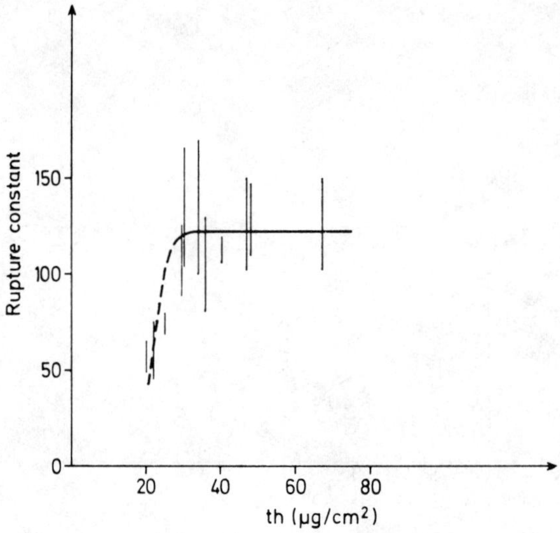

Figure 4. Rupture constant of polyimide foils as a function of foil thickness.

Table I: Comparison of Rupture Constants of Different Foils

	g/cm^3*	Chen(2)	Spivack(3)	CBNM-Values
Vyns	1.39	-	40	54 ± 8
Formvar	1.23	102	102	100 ± 9
Carbon	-	-	-	10 ± 4
Polyimide	-	-	-	122 ± 21
Mylar	1.40	-	233	-
Polycarbonate	1.20	-	173	-
Parylene C	1.29	-	71	59 ± 11
Parylene M	-	-	64	-

*Specific masses used(4) to recalculate the values of Spivack.

clearly illustrated by the results of pressure - dropping tests carried out as described in ref. (1) (Figures 2 and 3). At thicknesses below 30 $\mu g/cm^2$, there is however still a problem which is illustrated by the fact that the rupture constant is no longer a constant but a fastly decreasing value as is shown in Figure 4. Experience in target preparation by both electrospraying and vacuum evaporation confirm that this is the limit of good quality.

The mechanical strength of the improved polyimide is comparable to that of formvar and significantly better than that of the other foils used in target preparation (Vyns, Parylene, Carbon).

The comparison of rupture constants determined in other laboratories(2,3) and at CBNM, shows that our present polyimide foils seem to belong to the strongest ones (cfr. Table I).

The only "stronger" materials - according to Spivack(3) - would be mylar (polyethylene terephtalate) and polycarbonate, but these materials only seem to be available at thicknesses of about 500 and 250 $\mu g/cm^2$ respectively(4).

3. Aging

Aging of polyimide foils was studied, because it is important to be able to store the foils before target production and to use the targets over relatively long periods afterwards.

The storage of amide-acid solutions was found to be impossible under normal laboratory conditions. After a period of 2 months of conservation the viscosity of the solutions had decreased considerably and a slight precipitation could be observed. Foils prepared from these solutions were no longer transparent when they were on the glass plates, and their later release proved to be impossible.

On the contrary, foils stored on glass plates or self-supported foils packed in metallic boxes did not show any significant decrease in mechanical strength, even after a 7 months period. Furthermore, the production yield of the foils stored on glass plates remained essentially the same as the one of freshly prepared foils.

Although this was not examined systematically, it must nevertheless be stressed that unpacked foils stored in the laboratory environment became very brittle after some months and even completely disappeared when they were subject to a slight airflow.

V. RESISTANCE TO RADIATION

The radiation resistance of different types of foils to 6 MeV alpha-particles is presently studied using a cyclotron beam. The results of this study will be published later on.

However, first results show that compared to Vyns, Formvar resists only slightly better, while Parylene C and Polyimide are resp. 50 and more than 100 times better. Two polyimide foils of resp. 22 and 34 $\mu g/cm^2$ were even not destroyed after an integrated dose of 7×10^{16} alpha-particles per cm^2.

VI. CONCLUSION

The modified preparation method allows the production of polyimide foils with significantly better mechanical strength. This is shown by both pressure and dropping tests. Furthermore, their production yield is practically quantitative, at least for thicknesses above 30 $\mu g/cm^2$. Compared to other plastic foils, the resistance to chemicals, heat and radiation is excellent. Problems remain however when foil thicknesses below 30 $\mu g/cm^2$ are needed, and the real thicknesses of the foils is difficult to foresee especially in the range of 20 to 40 $\mu g/cm^2$. The foils can be stored for relatively long periods in metallic boxes or on glass plates without significant deterioration.

REFERENCES

(1) J. Pauwels, J. Van Craen, J. Van Gestel, and J. Van Audenhove, Polyimide Substrate Foils for Nuclear Targets, INTDS - Conference, München, 11-14 Sept. 1978.
(2) J. J. L. Chen, Test of Rupture Strength of Thin Polymer Films, Rev. Sci. Instr. 21 (5), 491-492 (1950).
(3) M. A. Spivack, Mechanical Properties of Very Thin Polymer Films, Rev. Sci. Instr. 43, 7-12 (1972).
(4) M. A. Spivack, Parylene Thin Films for Radiation Applications, Rev. Sci. Instr. 41, 1614-1616 (1970).

ROLLING OF EVAPORATED MAGNESIUM ISOTOPES

Frank J. Karasek

Microfoils

ABSTRACT

The rolling of evaporated ^{24}Mg, ^{25}Mg, and ^{26}Mg foils from an initial thickness range of 200 to 300 μg/cm² down to an ultimate thickness of 60 μg/cm² or less is described. Roll pass reduction, pack thickness and surface conditions are the main factors in the successful preparation of these foils.

INTRODUCTION

Evaporated Mg foils, in the thickness range from 50 to 200 μg/cm², were almost impossible to remove from the substrate; therefore, it was decided to produce foils in this thickness range by rolling. In order to conserve time and effort, thicker foils were prepared by evaporation and these were rolled down to the desired thickness.

Rolling Procedure

Starting stock for the rolling operation were evaporated foils having a thickness range from 200 to 350 μg/cm². The foils were razor stripped from the glass substrate and as a result had a great tendency to curl up into a tightly wound cylinder.

The stripped foil was inserted in a stainless steel pack and this assembly was cold rolled several passes. Reductions were on the order of 5% after every second roll pass. After achieving a total reduction of approximately 20%, rolling was discontinued and the foil was removed from the pack. The main purpose of this initial rolling was to flatten the foils so that it could be cut into sections more suitable for rolling and weighed so as to determine the thickness.

The cut and weighed foil was then inserted in a new stainless steel pack and rolled, again using reductions of 5% after every double pass.

The foil was removed after each 50% total reduction and was reinserted in a new pack. As rolling progressed and foil decreased in thickness to a range from 100 to 150 $\mu g/cm^2$ (estimated) the stainless steel packs were renewed after every 20 to 30% total reduction.

After the approximate foil thickness was reached (as estimated from the area increase), the foil was removed from the pack and placed between weighing paper. The foil was sectioned to the size required for mounting on the target frame.

The above rolling method was used successfully for the preparation of ^{24}Mg, ^{25}Mg, and ^{26}Mg foils at thicknesses of 125, 100, 75, 60, and 50 $\mu g/cm^2$. Sufficient area was generated at each thickness level so that several targets were available from each of the rolled foils.

The stainless steel packs used during this rolling operation were nominally only one half the thickness that has been used in other foil rolling operations. The use of a thinner rolling pack, multiple rolling passes and frequent pack changes were a great help in preventing the Mg foils from sticking in the pack. The generation of excessively large foil areas and for foils with an excessively long single direction was also avoided so as to prevent destruction of the foil by splitting and/or spot sticking.

CERAMIC AND CERMET TARGETS*

E. H. Kobisk, T. C. Quinby, and W. S. Aaron

Solid State Division
Oak Ridge National Laboratory
Post Office Box X
Oak Ridge, Tennessee 37830

ABSTRACT

Use of isotopic materials as targets in high temperature environments, e.g., reactor cores, requires that chemically stable forms of the isotopes be employed. Usually oxides are compatible with temperatures >1600 K, although some light element oxides exhibit some volatility at temperatures >1300 K. Especially in the case of heavy elements, the relatively low melting points of the metals, poor compatibility of the metals with encapsulation materials, and high chemical reactivity at moderate temperatures preclude the use of metal targets. However, encapsulation of ceramic targets has been successfully performed yielding high integrity samples. If hydrogen-reducible metals are mixed with the isotope(s), malleable, high strength, corrosion resistant targets can be rolled which contain a ceramic phase of isotope oxide. Isotope dilutions, additions of metals to form the metal matrix of a cermet target, and subsequent homogenization of all components are performed by dissolution in molten urea followed by calcination and compaction into the desired target form.

INTRODUCTION

For more than 20 years, Oak Ridge National Laboratory

*Research sponsored by the Division of Waste Products, U. S. Department of Energy, under contract W-7405-eng-26 with the Union Carbide Corporation.

personnel have prepared unique and exacting specimens of separated stable and radioactive nuclides in physical forms compatible with research in physics, chemistry, and metallurgy. Although not always used for studying subatomic particle interactions with specific nuclei, most samples have been used in this manner by researchers throughout the world. Isotope samples in "target" form can assume physical shapes such as thin films, wires, rods, castings, spheres, and many others. To shape these target samples a variety of methods are employed. However, it is not always possible to prepare the required target either because insufficient material is available, or reduction of a compound form of the nuclide to the element is not feasible, or the end product would not be physically and/or chemically compatible with the temperature or other materials to be used in the desired experiment.

To circumvent some of the above described problems, a technique has been developed to prepare shaped ceramics or cermets (metal-ceramic mixtures) which have the ability to withstand enhanced temperature regions and yet have readily definable composition and high uniformity of nuclidic distribution. Ceramic materials are, for the most part, more compatible with encapsulation materials than are elemental forms and almost always have lower vapor pressures and better chemical stability so that their use in vacuum or under oxidizing conditions is permissible.

This paper describes methods of preparation of ceramic and cermet starting materials for target preparation, methods of compaction and forming, and the characteristics of resultant specimens.

Experimental Development

In 1971 W. D. Box reported at the Third International Symposium on Research Materials and Nuclear Measurements on the preparation of neutron dosimetry samples usable in reactor core regions for the purpose of measuring neutron flux and energy distributions. In this work, powdered oxides of a variety of separated stable and radioactive nuclides were encapsulated in high purity vanadium containers having an outer diameter of 1.25 mm and lengths varying from 3 mm to about 10 mm. Each capsule was uniquely identifiable and contained well-analyzed material, the quantity of which was determined by weight-by-difference on a microbalance. Masses of contained material per capsule varied from 0.1 mg to 9 mg with a measured error of ±1%. Preparation of these dosimeters was tedious and subject to error resulting in frequent sample rejection because of the inherent difficulty in handling the powdered materials. Closure welding, decontamination and weighing-by-difference measurements all resulted in time-consuming and error-producing events which subsequently caused the unit cost of acceptable samples to be high.

Chemical Treatment

Recent developments have permitted preparation of similar samples at a cost approaching 50% of the earlier costs. A method whereby desired nuclides could be prepared as easily formed and sinterable powders has been achieved in which the nuclide of interest is dissolved and subsequently precipitated from a molten urea solution (450 K). Dilutions of the nuclide of interest with inert materials, e.g., 1% $^{239}PuO_2$ in MgO, have been prepared in this manner as well as pure materials.

The process simply involves quantitative dissolution of materials containing the nuclides of interest in nitric acid (or directly in molten urea when possible) and addition of urea so that the urea-to-nitrate concentration ratio exceeds 5:1. In batch processing this ratio must be exceeded to avoid severe ebullation during the precipitation process and the subsequent thermal decomposition of excess urea. After precipitation is complete, excess urea is decomposed at temperatures >450 K and eventual calcination of the material at approximately 1110 K in air quantitatively produces a uniform particle size powder.

Depending upon the urea-to-nitrate concentration ratio, particle size can be controlled in the range of 0.05 μm diam. to 10 μm diam., the smaller diameter being formed from solutions containing the larger urea concentration. If easily sinterable materials are desired, the particle size should be small. For example, thoria (ThO_2) whose melting point is ∿3500 K, has been sintered into a translucent body of approximately 98% theoretical density at 1720 K when the ceramic powder was produced by precipitation from a 20:1 mole ratio of urea to nitrate ion.

Of importance is the ease with which chemical composition of the oxide powders can be controlled and the subsequent shaping and densification into the desired physical form is achieved. Because the "urea process" is quantitative, dissolution of carefully compounded admixtures of materials at the beginning of the process result directly in the same composition of the precipitated powder in the end, the only possible variation arising from volatility of a component at the required calcination or sintering temperature. Dilutions of isotopically pure materials in MgO and Al_2O_3 have been prepared in which variation of composition did not occur to more than ±0.1% after the physical form was produced.

Figure 1. Green ceramic wire extruded from a steel die using 2% polyethylene in paraffin as the binder-lubricant.

Physical Forming Procedures

Depending on the physical shape of the target sample, a variety of forming and densification procedures can be used. In the case of neutron dosimetry materials noted above, a "wire" form of the ceramic containing the nuclide(s) of interest is most

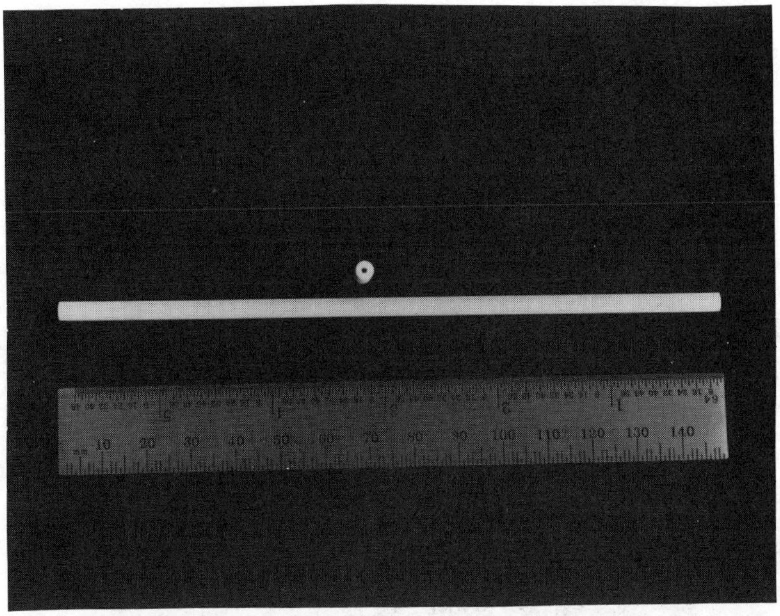

Figure 2. Magnesia tubes are used to contain green extruded ceramic wire so as to maintain straightness during sintering.

desirable. By mixing the calcined powder prepared from the urea process with a lubricant-binder material, e.g., 2% polyethylene in paraffin, wire can be extruded, Figure 1. Subsequent heating of the formed material in MgO tubes, Figure 2, causes the powder to sinter (the binder being thermally volatilized and/or chemically decomposed) with resulting densities >90% of the theoretical value. High-fired MgO tubes are used to contain the "green" extruded material during the sintering process so that straightness of the wire is maintained.

Figure 3. Continuous extrusion of green wire and subsequent spooling permits up to 100 m lengths to be formed.

A system for continuous extrusion of green material is shown in Figure 3 wherein wire is spooled on a reel after extrusion. As much as 100 m of wire has been produced by this latter technique with a single loading of the extrusion die. Typical dosimeter

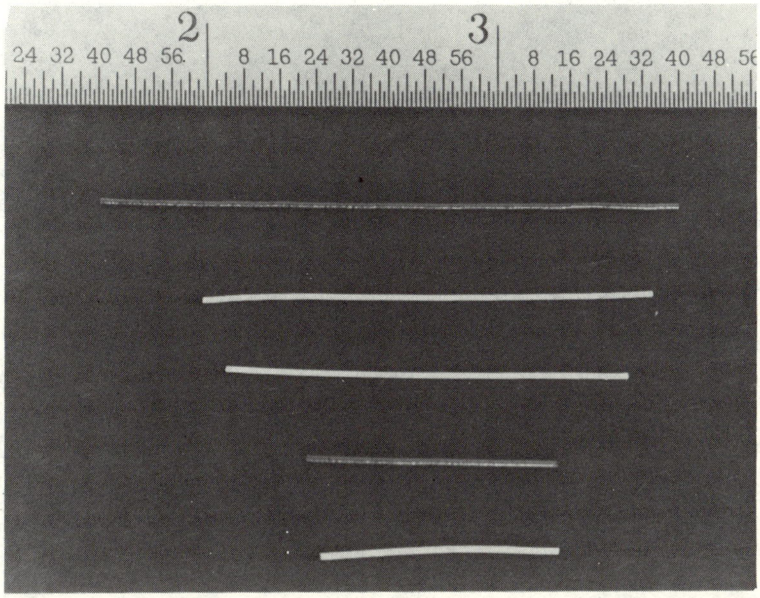

Figure 4. Typical dosimeter wires as formed by extrusion and sintering. From top to bottom: UO_2, 1% CoO-MgO, Sc_2O_3, Mn, 13% NiO-MgO.

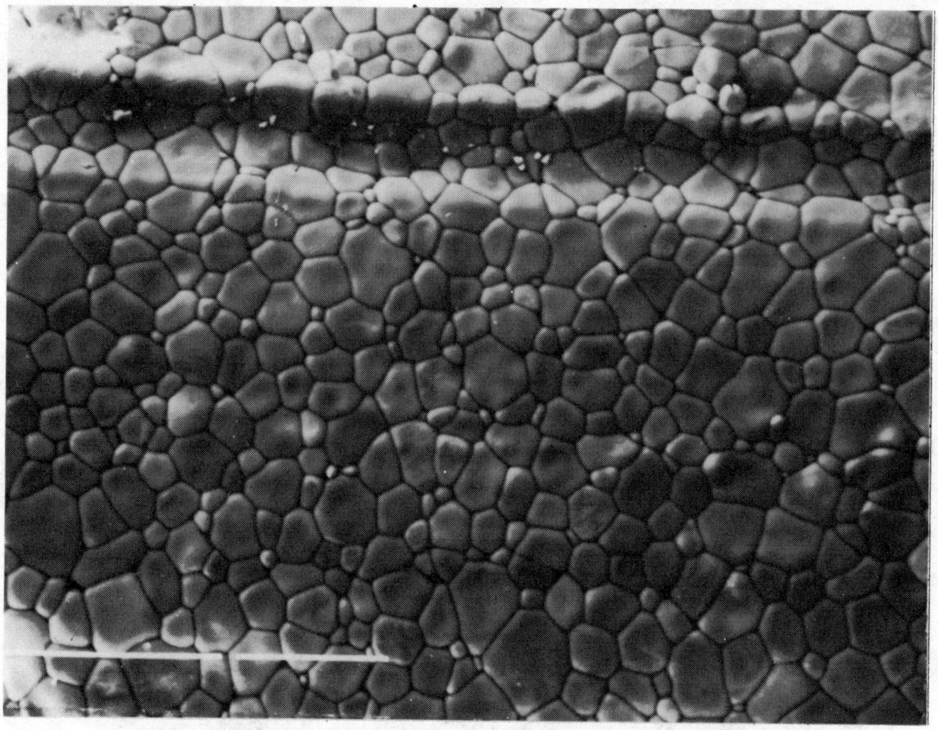

Figure 5. Surface structure of a ThO_2 extruded and sintered wire illustrates high density structure. Note extrusion mark at top of photomicrograph (9000X).

wires produced by this procedure are shown in Figure 4. A photomicrograph of sintered thoria is shown in Figure 5 to illustrate the high density structure achieved by the sintering of urea-produced ceramic powder.

The efficacy of extrusion of sinterable isotope-bearing ceramics is dramatically illustrated by our preparation of $^{244}Cm_2O_3$

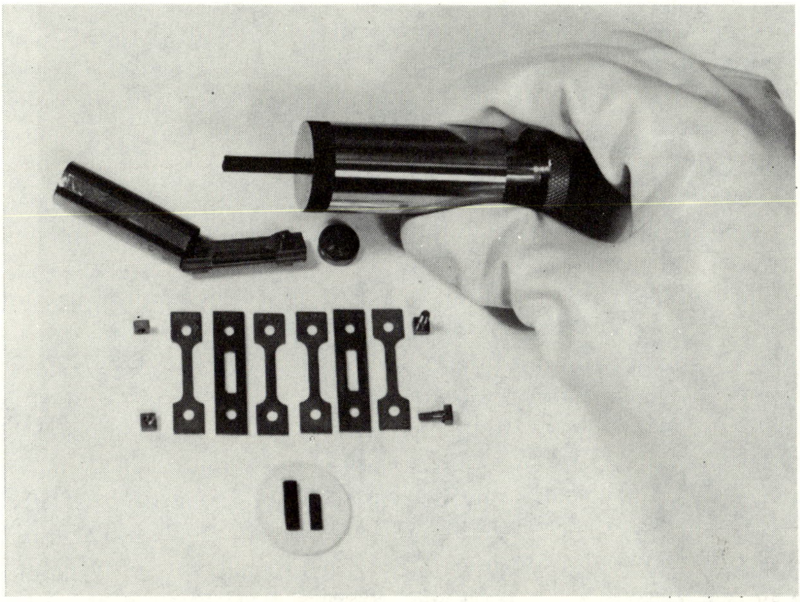

Figure 6. Cm_2O_3 extrusion with a wax binder produced ribbons suitable for sintering. Illustrated is the extrusion die, green and sintered material, and sample holder assemblies. Ribbon samples before and after sintering are shown at the bottom of the photograph.

ribbon of rectangular cross-section. Lengths of 21 mm of 3.0 mm x 6 mm cross-section Cm_2O_3 were required for material compatibility and radiation damage studies. By extrusion using a Roger A. Reed Corp. wax (known as No. 77030) as the binder and subsequent sintering these samples were prepared. Resultant density was 59.4% of the theoretical value. Since samples were required to be fitted

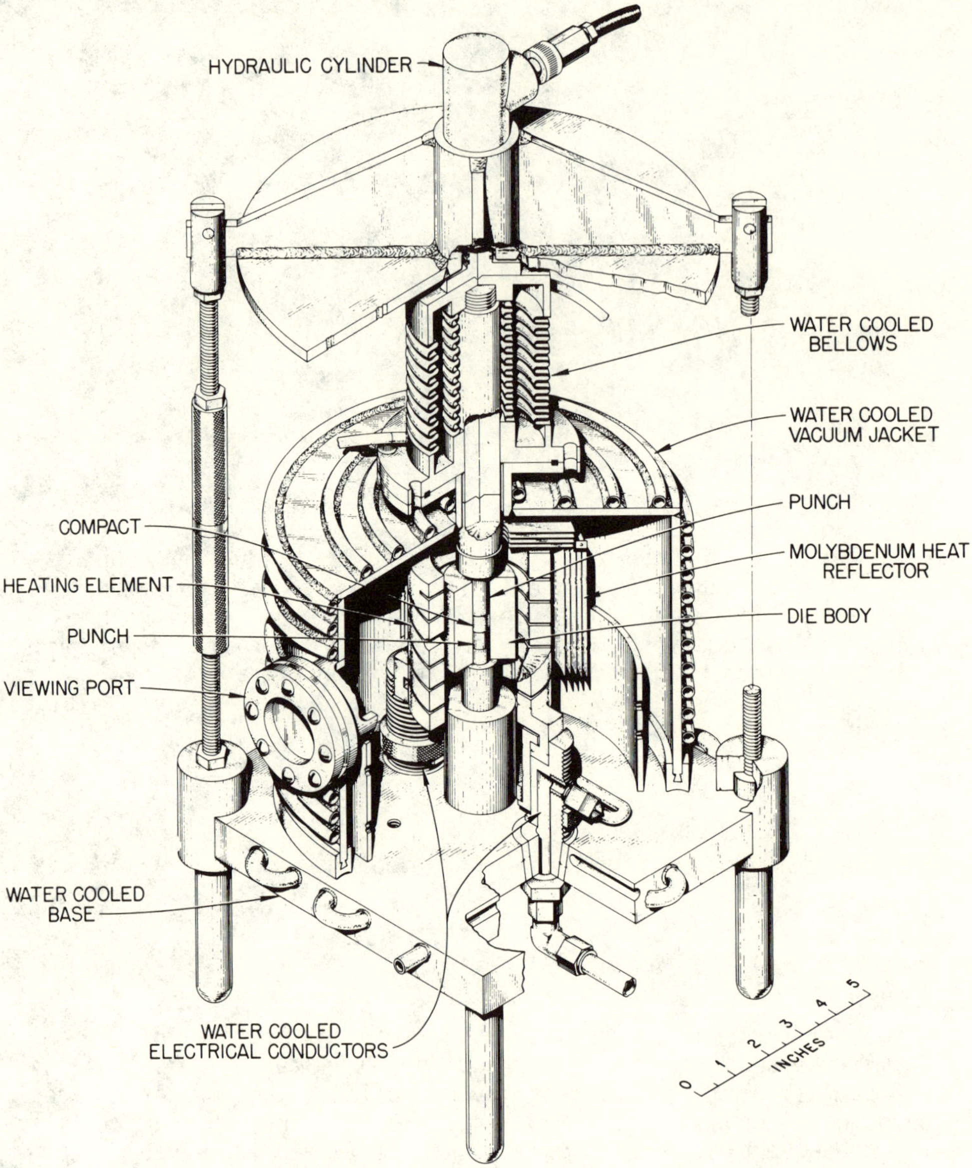

Figure 7. Hot press schematic diagram illustrates simplicity of preparation of flat, disk-shaped actinide oxide samples.

Figure 8. A disassembled view of thermal pressure bonding die is shown in photograph. At lower center-right is shown a ceramic, hot-pressed disk of UO_2 to be encapsulated in the copper shell on which it rests. Larger copper capsules are then bonded around the small inner capsule (No. 41 shown in illustration).

Figure 9. A pressure bonding die assembly in the closed position is shown; assembly used to form large capsule enclosures is illustrated in Figure 7. High current-low voltage power is used to heat the capsule parts under pressure by resistive heating. Outer sleeve (middle of photograph) is used to center capsule components and to shear expanded metal to appropriate diameter after press-bonding.

to +0.0 mm, -0.05mm, in terms of thickness, width, and length, the extrusion die nozzle dimensions were defined after determining shrinkage of the extruded material upon sintering at 1700 K. The extrusion die assembly, extruded and sintered samples, and the sample holder is shown in Figure 6. All operations with this material were performed in an argon atmosphere glove box to maintain the sesquioxide composition. Heat generated by the ^{244}Cm decay required the use of a high melting point wax binder (noted above) and extrusion and sintering were required to be performed within 12 hr. because radiation degradation of the wax precluded satisfactory forming if longer periods were involved.

Another forming method is hot pressing which has been employed to produce disc-shaped samples of actinide oxides as used in neutron reaction rate measurements. Powders produced by the urea process are easily densified in ATJ graphite dies by hot pressing at temperatures of 1100-2000 K and at pressures of 14-34 MPa (2000-5000 psi). A schematic representation of a typical hot press used for this purpose is illustrated in Figure 7.

The hot pressing procedure is used to form rods of an actinide oxide which can be sawed, subsequent to pressing, using an abrasive wheel or a diamond saw. Densities approaching 98% of theoretical have been achieved and circular wafers of approximately 4 mm thickness have been produced from actinide oxides including americium. Because of the hazardous nature of these materials, such samples must be encapsulated for handling and/or insertion into a reactor core zone. To encapsulate disc-shaped samples a thermal pressure bonding technique was developed. A typical pressure bonding system is shown in Figures 8 and 9. Copper or aluminum has been used to encapsulate the isotope samples by this technique.

The pressure bonding apparatus is simply a circular die prepared from tungsten carbide which will not deform significantly under pressures of 14-28 mPa (2000-4000 psi) and temperatures of approximately 750 K. Preformed discs of encapsulating metal, Figure 8, are inserted and the ceramic actinide disc is centered in the metal sandwich. Subsequent pressure and heating by electrical resistance in the die at the periphery of the capsule results in a bond which, when examined metallographically, shows no definable interface between the metal layers. The outer sleeve surrounding the ram of the die is used both to center the composite pieces in the die and to subsequently shear the finished capsule to the appropriate diameter.

Results

Use of urea-produced oxide powders permit preparation of

Table 1. Varieties of Extruded-Sintered Ceramic Wires Produced by the Isotope Research Materials Laboratory

Composition	Diameter (mm)	Length, m
$^{232}ThO_2$	0.5, 0.7	350
$^{233}UO_2$	0.5, 0.7	350
$^{235}UO_2$	0.5, 0.7	350
$^{238}UO_2$	0.5, 0.7	350
$^{237}NpO_2$	0.5, 0.7	350
$^{239}PuO_2$	0.5, 0.7	350
Sc_2O_3	0.5, 0.7	350
1% $^{233}UO_2$-MgO	0.5, 0.7	300
1% $^{235}UO_2$-MgO	0.5, 0.7	610
1% $^{238}UO_2$-MgO	0.5, 0.7	610
1% $^{237}NpO_2$-MgO	0.5, 0.7	610
1% $^{239}PuO_2$-MgO	0.5, 0.7	610
1% $^{232}ThO_2$-MgO	0.5, 0.7	610
16% $^{61}NiO_2$-MgO	0.8, 1.2	50
1% CoO-MgO	0.5, 0.7	610
$^{238}UO_2$	4.8	0.5
$^{238}UO_2$ (Tubing)	4.8 o.d., 2.8 i.d.	0.5

unique ceramic and cermet target samples. Dilutions of specific isotope nuclei can be produced by addition of the diluent nuclide in the dissolution step of the urea process which results in essentially atomic homogenization of the mixed species. By this preparative process, a wide variety of pure and diluted isotope samples have been prepared, Table 1.

Extrusion of coprecipitated and calcined oxides can be performed to shape materials into desired forms having very narrow dimensional tolerances. Table 1 notes the variety of shapes and sizes that have been formed by extrusion. Preparation of samples of this type would be impossible by other methods, especially where highly radioactive and toxic materials are required. The urea process and subsequent forming processes are readily compatible with a glove box environment and have been used in this mode at ORNL.

If urea-precipitated powder contains a reducible cation(s) species, subsequent heating of the species in a reducing atmosphére yields a metal phase of uniform distribution on a nearly atomic scale in the mixture. Subsequent compaction and densification of such a mixture results in a cermet in which the ceramic phase is microencapsulated in a metallic phase. Composition of the metal and ceramic phases can be tailored by additions at the beginning of the urea process.

Depending upon the quantity and composition of the metal phase, it is possible to form cermets (and alloys) which are malleable and formable by standard rolling and forging techniques. For example, actinide oxide in a copper or nickel matrix can be formed by rolling into highly uniform foils. Provided the concentration of the ceramic phase is insufficient to harden the matrix metal to a nonmalleable condition, thin foils can be rolled nearly to thicknesses achievable with the pure metal. Distribution uniformity of >99% have been achieved in this manner. Cermets are amenable to cold forming under pressure and excellent samples suitable for cross-section measurements can be prepared in this manner.

Finally, it should be recognized that this technique is applicable to preparation of thick samples of many difficult to handle isotopes, e.g., Ru, Ir. Dispersion of these elements in metal matrices is sometimes the only method of achieving a desired sample thickness or configuration. Unusual shapes can be prepared from urea-precipitated materials. A specific case in point was the preparation of 12-mm diam. UO_2 tubing with a 2-mm wall thickness. This material was prepared by extrusion in tube form using a wax binder. Because of shrinkage upon sintering, the green extruded samples were about twice the diameter of the final densified specimens.

Conclusions

The urea precipitation process together with a variety of physical forming methods adds considerable versatility to target and other form preparation capability. The technique is especially adaptable to the forming of sinterable ceramic samples and offers an alternative preparative method for some hard-to-handle materials. Preparative processes are amenable to glove box operations and are very useful in producing high quality actinide samples. Cermets containing desired nuclei in the ceramic phase can be cold-compacted and even formed by rolling and forging. Although not a panacea for all target preparative problems, ceramic and cermet target forms may be the only applicable specimen forms suitable for the experimental environment in which they are to be used. The relatively simple thermal pressure bonding method of sample encapsulation offers an additional dimension of versatility in providing leak-tight, reactor compatible samples.

DRY SETTLING AND PRESSING AT IUCF*

William R. Lozowski and Tina M. Rife

Indiana University Cyclotron Facility

ABSTRACT

A variety of pressed powder targets of 2-cm dia. have been produced using a compact, simple, and material-conservative technique. Although the attainable thicknesses are highly specific to the material and experiment, the basic method, and variations of it, have been useful in making very thin binder-supported targets, two to twenty-five mg/cm^2 unsupported targets and very thick conventionally pressed targets. Rudiments of the method are: 1. two mechanical vibrators linked by a wire loop, 2. a narrow glass tube capped with an appropriately sized stainless steel screen, 3. vacuum evaporated carbon-coated glass slide pieces as the substrate for the settling and pressing operations, and 4. counter rotation of the substrate and sifting tube during settling.

INTRODUCTION

Often the first step in producing targets for IUCF is distribution of an isotopically enriched powder within acceptable uniformity limits of ± 10-15% over a circular area of 2 cm diameter. The dry settling procedure to be described uses several components which may be varied independently to distribute powders with diverse physical properties. Most frequently it is used to distribute 3-25 mg/cm^2 of a powder onto a 20-50 μg/cm^2 evaporated carbon-coated glass slide piece(1). A like slide piece is care-

*Work supported in part by the National Science Foundation.

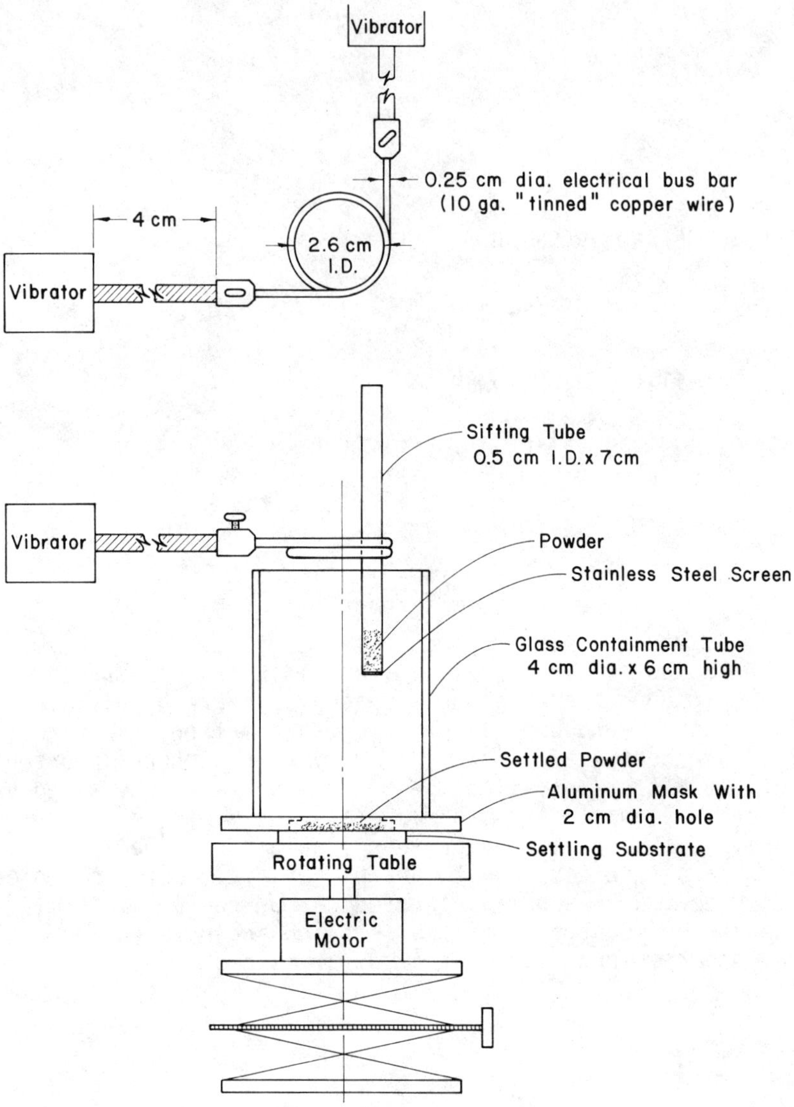

Figure 1.

fully positioned on the powder-coated one and the sandwich is pressed to obtain a self-supporting target. This procedure has, however, been used to distribute Re powder uniformly into a benzene/styrene solution. It has also been used to settle powders into a pellet press for targets of 50 mg/cm^2.

Although the dimensions of the components are interrelated, they are not critically so.

For air or moisture sensitive materials, it should be possible to set up the apparatus in a controlled atmosphere.

Areal density measurements of small razor-cut sections of sacrificial targets are usually used to determine targets uniformity. However unpalatable willful target destruction may be, areal density measurements can often be made quickly and with very little loss of material.

Targets made with this method include ^{Nat}S, ^{Nat}C, ^{12}C, ^{13}C, $^{151}Eu_2O_3$, $^{152}Eu_2O_3$, $^{148}Sm_2O_3$, $^{154}Sm_2O_3$, $^{183}W\,O_3$, $^{186}W\,O_3$, $Ca\,H\,PO_4$, and Melamine[2,3].

General Procedure

As one can determine from Figure 1, the preparative set up is quite simple. Two mechanical vibrators with individual amplitude controls are linked by a 2.5 mm dia. wire bent into a circular loop of 26 mm I. D. The vibrators are adjusted relative to each other by observing (from overhead) the visually blurred area around the oscillating ring. When this area is as circular as possible, the vibrators are correctly adjusted.

The sifting tube consists of a cut-off Pasteur pipet of 5 mm I. D. with a stainless steel wire screen epoxied to one end. Screens of mesh numbers 150, 250 and 325 are most often used. After the tube is loaded with powder, it is hand held and rotated around the inside surface of the vibrating loop at 100-240 rev/min. This produces a shower of powder which may be examined through the wall of the glass containment tube with the beam from a directed light source. Such a light source is also necessary to detect powder particles floating out of the apparatus. The sifting tube may be lowered 3-4 cm into the 6-cm high containment tube to minimize this loss. If excessive powder loss is encountered, the length of the sifting and containment tubes may be increased.

A mask with a 2-cm dia. hole is used to define the target area. When the amount of powder available to make a target is severely limited, the sifting tube may be reloaded as needed, from the "overspray" on the mask. Removal and replacement of the mask is easy if the underside has been cut back somewhat deeper and larger than the settled powder.

The containment ring, mask and settling substrate are rotated at 6 rev/min., in the opposite direction of the sifting tube rotation, by a small electric gear motor. This assembly is supported by a small lab jack and can be moved as a unit close to or away from the wire loop as desired.

Discussion

1. The Vibrators

Both vibrators have maximum linear motions of 5 mm and output shafts of 6.3 mm dia. glass-reinforced thermosetting resin. The flexibility of these shafts enables one to adjust the vibrators so that the oscillating wire loop does not have motion nodes. The wire loop was made by bending a # 10 "tinned" copper bus bar around a mandrel.

One of our vibrators is a discontinued Corning[R] model LM-2 solution stirrer. The other was made in the lab from the coil of an electric solenoid valve. If one is tempted to construct such a device, two design considerations are offered: a hollow steel plunger must be used to avoid overheating the coil and a diode must be used to provide a pulsed voltage to the coil. The coil plunger protrudes from the rear of the coil and is held in place by a spring. When the coil is activated, the plunger is drawn forward, deeper into the coil. When deactivated, the spring pulls the plunger back.

2. Sifting

When possible, one loads the sifting tube with more powder than is required to make the target. This is desirable because in practice it is difficult to get the entire amount of powder through the screen and onto the desired area. Sifting a powder is sometimes delightfully trivial (e.g., with C and Melamine), but often problems occur (e.g., with S and WO_3). Some powders will clump in the settling tube when vibrated. Still others will clump even as they fall through the containment tube. Drying the powder and sifting tube thoroughly is often of tremendous help. Another possible remedy involves sifting the powder with the aid of small metal strips or ball bearings in the sifting tube. Deposits from powder sifted in this way are usually non-uniform. But if the powder is collected, reloaded into the settling tube and redried, an acceptable settling rate may result. For marginally acceptable rates (i.e., greater than five minutes) a lab jack is useful to support one's stirring arm.

The variable parameters of settling are:

a) the diameter of the wire loop
b) the intensity level of vibration
c) the grain size of the powder

d) the moisture content of the powder
e) the size of the screen
f) the depth of the sifting tube in the containment tube
g) the rotation speed of the sifting tube
h) the counter rotation speed of the settling substrate.

Only a few comments on these parameters are necessary. If the center of a trial target is too thick, the wire loop may be made larger, the vibration level reduced, the screen size changed or the depth of the sifting tube in the containment tube changed. Holding the sifting tube higher in the containment tube will cause less stirring of the air; however, one must be careful that powder does not float out and lost.

The mask hole size is, of course, also variable subject to target area requirements. In many instances, a target of less than 1-cm dia. can be used at IUCF. But instead of reducing the mask hole size for a thin, likely-to-break target, we often retain 2-cm dia. and trim a usable piece from unbroken areas after pressing.

3. Substrates

Evaporated carbon-coated slide pieces are normally used as the substrate for settling and pressing because the adhesion of pressed powders to them is less than for any other surface we have tried[(4)]. Most pressed targets below a thickness of 20 mg/cm^2 will form a bond to press surfaces which is stronger than the inter-particle bonds formed by pressing. Hence the surfaces in contact with the powder must be exceptionally non-sticky (as well as hard) to achieve the thinnest possible self-supporting target.

A conventional pellet press is used with this settling technique when the settled powder is thick enough to require confinement on the perimeter but too thin to dump and spread it routinely into a press cavity. Sticking problems with these thicker targets (25-50 mg/cm^2) can often be solved by coating the press with a release agent which will not be a problem contaminant to the experimentalist. An attractive possibility is vacuum evaporating carbon onto the press platens. To date, however, we have not solved the problems of duplicating the non-stick quality of 20-50 μg/cm^2 carbon on glass with carbon films on hardened steel.

Liquid cells can be used to make supported powder targets which are too thin and/or brittle to be pressed and too heavy to suspend in the low viscosity solutions of greatest use. For instance, the Re target mentioned earlier was made by grinding ^{185}Re powder and settling it via the vibrating loop onto the surface of a 2-cm dia. Teflon^R cell containing a benzene/1 mg/cm^2 styrene solution. The Re powder--too rarefied at 4.5 mg/cm^2 to produce an unsupported target by pressing and too heavy to suspend in the solution long enough to settle uniformly--dropped undisturbed through the liquid cell. When the benzene evaporated, the cell was disassembled and the supported Re was floated from the glass slide cell bottom with distilled water.

4. Thickness Monitoring

To monitor the areal density of a settled powder on a preweighed carbon-coated slide piece, one has only to reweigh it. After pressing, we use a traveling microscope to obtain the final target area.

In situ determination of the amount of powder settled into a liquid cell or powder press is almost as simple. For these substrates, the loaded sifting tube is weighed before and after settling. The difference in weights minus the amount of powder recovered from the mask and the containment tube wall is the amount in the target area. Clearly, the accuracy of the estimation will suffer if powder is lost from the system.

5. Pressing

Pressing powders between carbon-coated slide pieces is a straightforward process. A slide piece with a defined area of settled powder is placed on a hardened parallel bar (5 cm wide x 2.5 cm high x 15 cm long). An old microscope, stripped of everything but the stage and adjusting screws and fitted with a vacuum pickup is used to lower the top slide piece into position for pressing. This is done carefully with the microscope fine adjustment knob.

The sandwich and parallel bar are then moved to the press. Neoprene rubber strips are placed beside the sandwich to support a second parallel bar (5 cm square x 2.25 cm high) over it. When pressure is applied, the rubber strips deform to allow the top bar to contact the sandwich gently. The maximum force that the glass slides will tolerate depends on the target uniformity and the parallelism of the press. Generally we use a force of 10 tons on a 2-cm dia. area.

Sticking problems are often a function of the pressing force and may vanish when it is changed. Oddly enough, stronger targets sometimes result when less force is applied.

Further Work

The described technique has evolved to the present state only within the past six months. To date we have not tried to sift binders into target powders, but it would seem to be an interesting direction of work. Perhaps another would be baking to increase the strength of pressed targets which are too fragile to be mounted and handled. Baking might also accomplish the release of a target stuck inside the sandwich after pressing.

One can say of this technique that there are enough things to "fiddle around with" while we wait for an eloquent solution to the problems of powders to be found.

REFERENCES

(1) Obtained from the Arizona Carbon Foil Co., 2239 East Kleindale Road, Tucson, AZ 85719, (602) 884-3495.

(2) ^{12}C and ^{13}C obtained from ProChem, 19 Ox Bow Ln., Summit, NJ 07901, (201) 273-0440. All other isotopes obtained from ORNL.

(3) This technique has also been used to prepare targets from environmental samples for elemental analysis.

(4) W. R. Lozowski, A Dry Powder Technique for the Preparation of Carbon Foils, Proceedings of the 6th Annual Conf. of the I. N. T. D. S., p. 115-119, Oct. 1977.

PREPARATION OF ^{14}C-TARGETS BY CRACKING OF $^{14}CH_3$-J

H. J. Maier

Sektion Physik der Universität München
8046 Garching, W. Germany

ABSTRACT

Thermal cracking of $^{14}CH_3$-J was utilized to prepare Ni-backed and self supporting ^{14}C-targets in the thickness range of 5-100 μg/cm^2. A detailed description of the method is given.

Table 1. ^{14}C Data

Properties of ^{14}C	
Type of radiation	β^-
$T_{1/2}$ (physical)	(5730±40)a
$E_{\beta max}$	156 keV
$E_{\beta\ av}$	50 keV
Specific activity	4.51 mCi/mg
$T_{1/2}$ (biological)	12 d
Radiotoxicity	Class IV (lowest)
Limit of free handling	10 μCi

1. INTRODUCTION

The problems, which are related to the preparation of ^{14}C-targets arise from two characteristics of this material, namely from its radioactivity and its high cost.

Some important ^{14}C-data are listed in Table 1. Although its radiotoxicity is low and its β-radiation is comparatively soft, law requires full radiation protection equipment for the handling of quantities larger than 10 μCi. To understand this 10 μCi limit in view of practical target handling, it is easily calculated from Table 1, that a target of thickness 10 μg/cm² has a specific activity of 45 μCi/cm², provided the enrichment is 100%.

The high cost of ^{14}C isotopic material requires an economic target preparation procedure. Therefore, electron beam evaporation, the usual method to prepare carbon foils, is ruled out. Instead of that the pyrolysis (or "cracking") of ^{14}C-labeled methyl iodide - was utilized. This latter method is known to have an efficiency of up to 30%(1).

Table 2. Bond Strength of some Organic Molecules(5)

Compound	Bond Strength kcal/Mol	Bond Strength kJ/Mol
CH_3-H	104	435
CH_3-F	108	452
CH_3-Cl	84	352
CH_3-Br	70	293
CH_3-J	56	234
H-CH_2J	103	431
CH_3-OH	91	381
CH_3-CH_3	88	368
CH_2=CH_2	172	720
CH≡CH	230	963

2. METHOD

The production of carbon thin film coatings by pyrolytical decomposition of suitable organic compounds is straightforward and has been used for target preparation in several laboratories(1-4). If a metal strip is heated to a temperature of 600-1500°C in an atmosphere of the compound in question the molecules may pick up a sufficient amount of energy to break their bonds and to be gradually decomposed into their elemental components. The result is, that finally a carbon deposit is formed on the metal strip. The efficiency of this procedure depends strongly on the bond strengths of the molecules, which have to be cracked. In Table 2, the bond strengths of a few relevant compounds are compiled. It is obvious that double- and triple-bonded molecules as for instance ethylene and acethylene with their high bond strengths will not work well in a thermal cracking procedure. Likewise, the pure hydrocarbons methane and ethane are relative strong bound. This fits well to recent reports(2,3), who state difficulties in preparing carbon foils by thermal cracking of methane. On the other hand, a continuous decrease of the carbon-halogen bond is observed in the series from methyl fluoride to methyl iodide. Thus it seems reasonable to use methyl iodide for the ^{14}C target production despite the possibility of a slight iodine contamination of the targets.

3. APPARATUS AND PROCEDURE

3.1 System Description

To understand the mode of operation of the cracking apparatus it is convenient to outline briefly the properties of methyl

Table 3. Methyl Iodide Data (for $^{12}CH_3$-J)(3,5)

CH_3-J = colorless liquid at 20°C

Molecular weight	141.95
Boiling point	42.4°C
Melting point	-66.5°C
Vapor pressure at 20°C	436.1 mbar
Vapor pressure at -196°C	<<1 mbar
Density (20°C)	2.28 g/cm^3

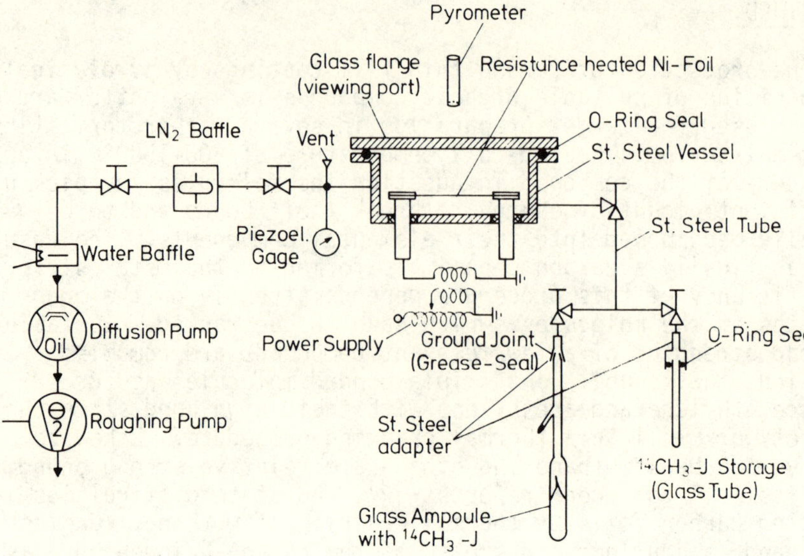

Figure 1. Schematic layout of the ^{14}C target preparation apparatus.

iodide. It can be seen from Table 3, that CH_3-J is a colorless liquid of density 2.28 g/cm^3 at room temperature. Its vapor pressure of 441.8 mbar at 20°C, respectively << 1 mbar at -196°C allows an easy transfer of the material from a storage cylinder into the evacuated cracking chamber and vice versa by gently heating the cylinder or immersing it in liquid nitrogen. The quantities of liquid $^{14}CH_3$-J one has to deal with are quite small. For example, 10 mCi or 2.22 mg of ^{14}C correspond to a volume of approximately 10 µl of liquid methyl iodide.

The cracking apparatus is shown schematically in Figure 1. The reaction chamber is fabricated of stainless steel and has a diameter of 10 cm and a height of 5 cm. It can be pumped to a vacuum of 10^{-6} mbar by a small diffusion pump to stack.

The carrier foil, which has to be coated with carbon, is mounted between two electrical feedthroughs which are connected to a low voltage power supply. Nickel foils which were supplied by the Goodfellow Company were used as carriers. Since the carbon is deposited on both sides of the heated foil, a large sheet with the dimension 65 x 40 mm is cut and then folded in its length directions, so that a double layer of 65 x 20 mm is formed. The open end is sealed by an additional narrow fold. At the end of the process the assembly is cut into two unilaterally coated strips.

The enriched methyl iodide is supplied in quantities of 200 mCi in hermetic sealed glass ampullas. These ampullas are not suited for a long-time storage of the material after breaking the sealing capillary, since they must be fitted to the system by means of a grease sealed ground joint, while methyl iodide strongly attacks all kinds of grease. Thus, before running the system, the methyl iodide stock is transferred into a previously evacuated, O-ring sealed vial by freezing with liquid nitrogen. In this type of vial, ^{14}C-enriched methyl iodide has been stored successfully over a period of several months.

Of course, the reaction chamber as well as the methyl iodide supply system is mounted in a glove box.

3.2 Operation

Running the system proceeds as follows:

1. The whole system is evacuated to a pressure of about 10^{-6} mbar. The methyl iodide is frozen with liquid nitrogen during this evacuation period.

2. The carrier foil is heated for a period of 1 minute in vacuum for reasons of surface cleaning.

3. A calculated amount of enriched methyl iodide vapor is allowed to enter the reaction chamber by gently warming up the storage vial. The quantity is controlled by a piezoelectric pressure gauge. A pressure of 20 mbar is sufficient to procedure a 20 $\mu g/cm^2$ thick carbon layer.

4. Cracking is performed by heating the foil to a temperature of 1000°C for a period of 3 minutes.

5. The residual methyl iodide as well as other gaseous cracking products are fed back into the storage vial by freezing with liquid nitrogen.

6. The target strip is cut into adequate sized pieces and - if required - the backing is removed by etching.

3.3 Etching

In the case of production of self-supporting targets, etching was performed in an acid bath composed of 1 volume part of concentrated nitric acid, 2 volume parts of concentrated acetic acid and 4 volume parts of distilled water. An adequate sized piece of foil is brought to swim on the surface of the etching solution, the Nickel foil being in contact with the liquid. After the Nickel backing is removed completely, the remaining carbon foil is washed by transferring it subsequently into three different water baths and is finally drawn on a target frame. This procedure works with a breakage rate of less than 10%.

The thickness of the Nickel carrier foil and the etching conditions had to be chosen according to the required target thickness. Very thin targets up to 10 μg/cm^2 were cracked on Nickel foils of 2 μm thickness. The etching speed had to be very slow for these thin foils, therefore the temperature of the etching bath was held at 20°C.

Thicker targets in the range of 20-30 μg/cm^2 were cracked on 12.5 μm thick Ni foils. Very thick carbon layers of 40-100 μg/cm^2 showed a bad adherence on the standard Ni carrier with its smooth surface. Therefore a thick Ni foil of 40 μm thickness was used which was roughened mechanically with emery cloth. In these two latter cases etching was performed with higher speed corresponding to a temperature of 40°C of the etching solution.

4. TARGET THICKNESS

The target thickness is a function of the methyl iodide pressure in the chamber, of the cracking temperature and of the operating time.

As to the temperature, we have found, that values significantly higher than 1000°C produce very brittle carbon foils with large internal stresses. So it is not possible to control the target thickness via the cracking temperature.

The dependence of the target thickness from the cracking time shows saturation character for fixed values of initial pressure and cracking temperature. It was found, that for the reaction chamber used at present a cracking time of three minutes was sufficient to complete the reaction.

In fact, target thickness was controlled by the amount of methyl iodide, which was allowed to enter the reaction chamber, observing a temperature of 1000°C and a cracking time of 3 minutes.

Table 4. Target Thickness as a Function of CH_3-J Pressuree in the Reaction Chamber

CH_3-J pressure mbar	5	10	20	50	100
Target thickness μg/cm^2	5±1	11± 2	18± 4	51± 5	85± 10

The thickness values obtained in this way are reproducible within an error range of ±20% and are compiled in Table 4. One reason for these relative large errors is probably the fact that it is not easy to maintain the foil temperature at 1000°C during operation.

5. IMPURITIES

Nickel and iodine contaminations of the cracked ^{14}C targets in the order of 1000 and 2000 ppm, respectively, have been found by Rutherford scattering as well as by x-ray fluorescence techniques.

6. CONCLUSIONS

A system has been set up which allows the routine preparation of nickel backed and free standing ^{14}C targets in the thickness range between 5 and 100 μg/cm^2.

REFERENCES

(1) G. C. Phillips and J. E. Richardson, Rev. Sci. Instr. 21, (1950), 885.
(2) R. Keller and H. H. Müller, Nucl. Instr. and Meth. 119, (1974), 321.
(3) F. Uihlein, Dissertation, Albert-Ludwigs-Universität Freiburg, 1962, (unpublished).
(4) E. Kashy, R. R. Perry, and J. R. Risser, Nucl. Instr. and Meth. 4 (1959), 167.
(5) Handbook of Chemistry and Physics (Ed. C. R. Waest, CRC Press, Cleveland, OHio, 1973-74).

PREPARATION OF SELF-SUPPORTING HOLMIUM TARGETS

Karl W. Scheu and Thomas Gee

Lawrence Berkeley Laboratory
University of California
Berkeley, CA 94720

ABSTRACT

A technique for producing thin self-supporting (150 to 500 μg/cm^2) ^{165}Ho targets on a 1 in. diameter target holder by vacuum evaporation using an electron bombarding source is described.

INTRODUCTION

Over a period of the last few years the Lawrence Berkeley Target Lab had received requests to prepare ^{165}Ho targets for use by experimentors at the Heavy Ion Linear Accelerator (HILAC). From the onset up to and including the present, we experienced difficulty obtaining a stable, self-supporting holmium target. In this paper we will describe the methods used to obtain these holmium targets, the thickness of which ranged from 150 to 500 μg/cm^2.

Vacuum System

The vacuum coater system used was designed and built by us at the Berkeley Laboratory. It is an all stainless steel system. The working volume of the vacuum chamber is 23 in. in diameter and 20 in. deep. The system has a 6-in. oil diffusion pump. A water cooled chevron baffle and a liquid nitrogen chevron baffle provide the trapping for the chamber. Rough pumping of the chamber is accomplished with a mechanical pump. The line connecting the pump to the chamber is equipped with a liquid nitrogen trap to prevent backstreaming of oil from the mechanical pump to the chamber. This system is capable of providing a base pressure of 8×10^{-7} torr and permits evaporation at 5×10^{-6} torr.

Source Preparation

The source used for evaporating holmium is an electron bombarding source. The crucible is made from tantalum 1/4 in. diam. and 1 in. long. The heater is a ring filament made from 40-mil diam. tantalum wire. Tungsten can also be used, but is slightly more difficult to form. A pre-firing of the source at approximately 2000°C in vacuum is performed to clean the tantalum crucible. The pre-firing of the source is very important to prevent contamination of the holmium metal used for evaporation.

Substrate Preparation

The substrates are cut from .00025 in. aluminum foil of 99% purity (stock material). The target substrate size is 1 in. x 1 in. Cleaning of the aluminum foil is performed by chemical treatment. The foil is placed in a diluted solution of sodium hydroxide (NaOH) for 2 min. After the 2 min. etching with NaOH, the aluminum foil is placed in a cold 70% nitric acid solution to minimize the effect of the subsequent alkaline cleaning. Then the aluminum foil is rinsed several times with deionized water. After the water rinsing the foil is rinsed with ethyl alcohol to remove the water. The aluminum foil is dried under a lamp.

We have also used evaporated aluminum foil (thickness range from 500 $\mu g/cm^2$ to 1 mg/cm^2) for the substrate of the holmium target. This produced a much smoother holmium foil.

Evaporation

The target substrate of aluminum foil is placed on a holder located 6 in. above the tantalum crucible. A glass slide is placed on top of the aluminum foil to prevent the foil from moving during the evaporation. A preweighed charge of holmium metal, in this case 225 mg of holmium, is placed into the tantalum crucible. The pressure in the chamber was lowered below 5×10^{-6} torr. It is essential that the pressure is below 5×10^{-6} torr before start of evaporation. The power input varies from 30 watts at the start of evaporation to 160 watts at the completion. At 5×10^{-6} torr, and 790°C the material will sublime. The total evaporation time was 2 1/2 min. During the evaporation process a blue-green plasma can be observed. This can be a good indicator as to when the holmium charge is exhausted.

Figure 1. Vacuum System used for Preparation of Holmium Targets.

Figure 2. Interior of Vacuum System with Set up for Holmium Evaporation.

Figure 3. Electron Bombarding Source.

Figure 4. Electron Bombarding Source with Graphite Heat Shield.

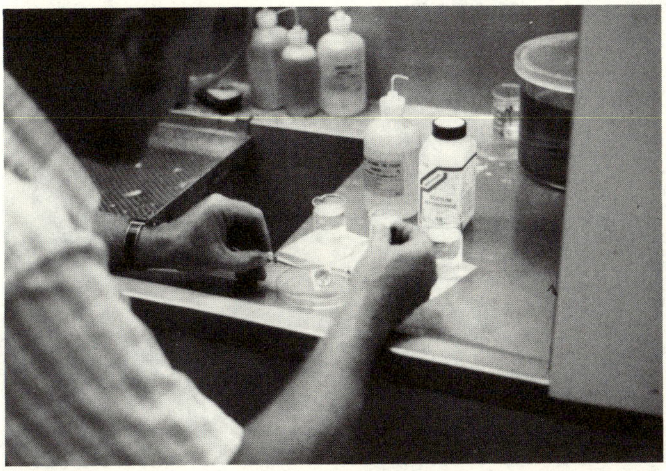

Figure 5. Preparing to Remove the Aluminum Foil from the Holmium Target.

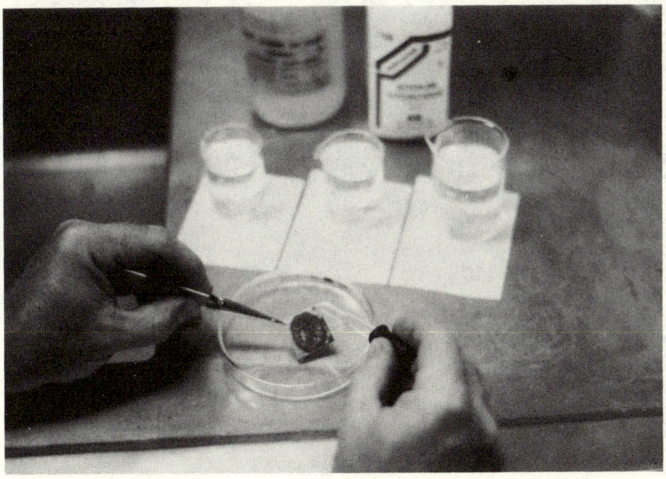

Figure 6. Close up View of the Removal of Aluminum Foil.

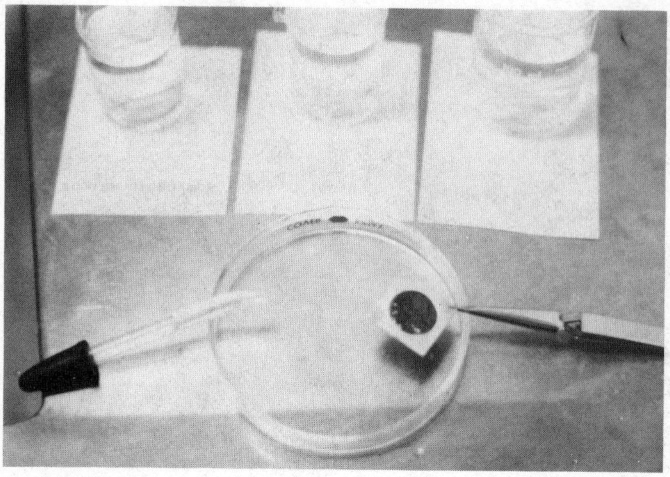

Figure 7. Holmium Target after Removal of the Aluminum Foil.

Figure 8. Holmium Targets.

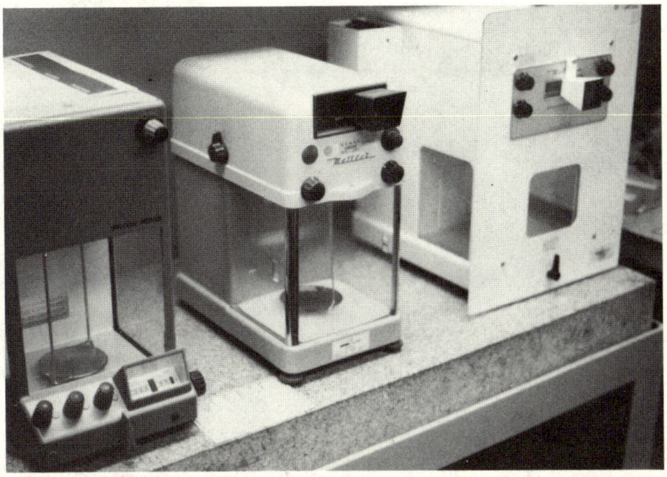

Figure 9. Mettler Microbalance used for Weighing the Holmium Targets.

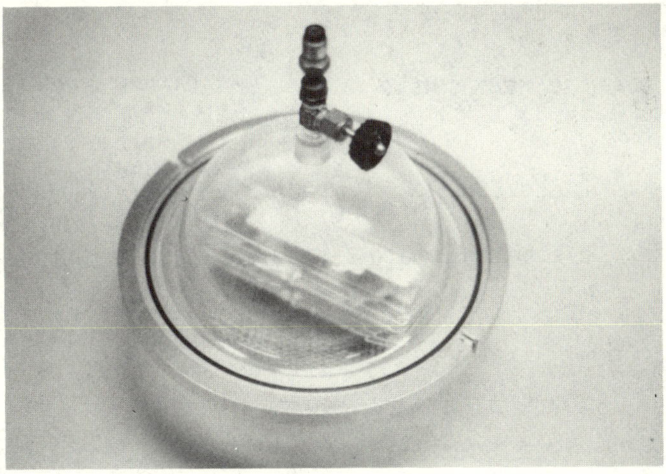

Figure 10. Vacuum Desiccator for Storage of the Holmium Targets.

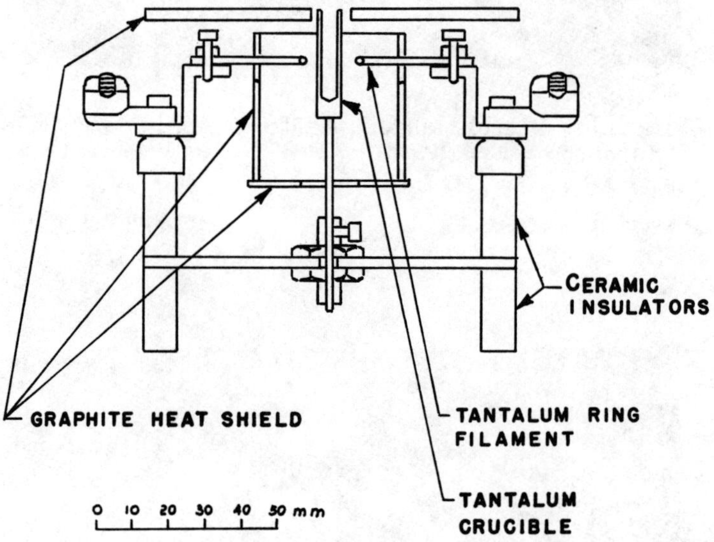

Figure 11. Electron Bombarding Source for Holmium Evaporation.

Foil Thickness Measurement

For the mass measurement of the foil we use a Mettler M5 microbalance, which has a precision of 1 microgram. The aluminum foil is weighed before evaporating the holmium. Then the aluminum foil with the condensed holmium is weighed again. Knowing the exact area of the holmium we can calculate the exact thickness of the vapor-deposited holmium.

Separating the Aluminum and Holmium

After we have determined the thickness of the holmium, we mount the aluminum and holmium sandwich onto a stainless steel holder. The holmium should be mounted so that it is in contact with the stainless steel holder, using epoxy cement. After the epoxy has dried, we start to remove the aluminum. The entire surface of the aluminum is carefully covered with a predetermined concentration of sodium hydroxide (NaOH) using an eye dropper. The NaOH should not produce a violent reaction when placed in contact with the aluminum backing. After several minutes one can observe the thinning of the aluminum. When the aluminum is completely etched away, no further reaction can be seen. The holmium foil that is mounted on the stainless steel holder is rinsed several times in deionized water and then rinsed with ethyl alcohol to remove the water. The foil is dried under a lamp. The holmium foil is stored in a vacuum desiccator until ready for use.

Target Condition and Conclusion

We have performed an X-ray fluorescence analysis of the holmium targets and have found small traces of aluminum and copper. These were very minute amounts of contaminent on the target and too small to cause any problems for the experimentors.

The strength of the target is very poor. Holmium is very brittle and great care should be taken in handling the target. We have experimented with different pressures during the evaporation of the holmium and have found that the lower the pressure, the less brittle the target. The percentage of success in making the holmium target is approximately 25%.

EIDGENÖSSISCHE TECHNISCHE HOCHSCHULE (SWITZERLAND)

D. Balzer
Nuclear Physics Laboratory
Hönggerberg, 8093 Zürich

Up to 1958 commercially available thick foils or thinner targets on a thick backing were used at the cyclotron.

The development of the fabrication techniques for thinner targets to be used with beams from the Tandem Van de Graaff accelerator started in the early 1960's:

-Fabrication of thin carbon foils by arc evaporation which was later substituted by electron gun heating for the evaporation. The foils were used directly in experiments, as a backing for other targets and as electron strippers in the accelerator.

-Use of an electron beam to heat a micro-crucible to evaporate enriched isotopes in quantities <1 mg.

In the period ∿1966-1975 one went through a peak of target-making activities. The laboratories concentrated mainly on evaporation techniques. A most probably incomplete list of various targets prepared:

-Self-supporting foils:

^{nat}C, ^{12}C, ^{13}C, Au, Ag, Al, Cu, ^{10}B, ^{11}B, Be, ^{50}Cr, ^{nat}Si; ^{6}Li, and ^{7}Li (prepared in a glove box and transferred in an air-lock facility).

-Foils on thin C or Au backing

In, Pb, ^{34}S, Sn, ^{47}Ti, ^{48}Ti, Th, rare earths.

-Sandwich foils:

Ca, ^{nat}Mg, ^{24}Mg, ^{26}Mg (between C or Au);
^{78}Se (between C); ^{238}U (between Ni and C).

From ∿1973 onwards thin foils of many elements with special emphasis on the purity of the targets were needed for atomic physics experiments.

From 1977/78 onwards investigations were started on the origins of contaminants in foils and on the fabrication of longer living carbon foils.

The future decline of nuclear physics experiments will make a search for new applications of the target-making technology necessary.

ROLLING OF SENSITIVE TARGET FOILS BEING COATED WITH EVAPORATED METAL LAYERS

Helmut Folger and Josef Klemm

Gesellschaft für Schwerionenforschung
Planckstr.1, D-61 Darmstadt, W. Germany

ABSTRACT

The rolling of metallic targets which are sensitive to oxygen and moist air can be facilitated by evaporating thin shielding layers onto both sides of the original materials. The obtained sandwich layers are mostly very resistant against destructive oxidations. They can be rolled down to required thicknesses even without the use of inert gas systems. The protected target materials are easy to handle, and they can be stored safely for rather long periods of time.

In this way U targets of thicknesses between 0.8 and 15.3 mg·cm^{-2} were prepared being protected with coverings of Ni of about 0.7% in weight of the corresponding U foils. Targets of Pr, ^{142}Nd, ^{143}Nd, ^{144}Nd, ^{146}Nd, and ^{148}Sm of a thickness of 3.5 mg·cm^{-2} were produced having coverings of Ti or Ni of 2.1 to 3.3% in weight of the target layers.

The preparation procedures for these kinds of metal sandwich layers are described. Some applications as targets in high intensity heavy ion bombardments with energies of up to 10 MeV/u of ^{48}Ti, ^{208}Pb, and ^{238}U ions at the UNILAC of the GSI are mentioned.

1. INTRODUCTION

Target substances which are easily attacked by oxygen and moist air as metallic uranium and a few lanthanide elements should be handled in glove-boxes filled with inert gases in order to avoid their partial or total destruction by oxidation. In some heavy ion bombardments with high energies the experimental results are not influenced by coating such sensitive materials with thin protective surface layers. Such sandwich targets are easy to handle, and they are especially useful in all cases where targets have to be transferred from one apparatus to another without the aid of inert gas systems. In addition to this, they can be stored safely for rather long periods of time in a desiccator or as well in a carefully closed plastic box containing an appropriate desiccant.

A procedure for the preparation of sandwich targets consisting of Ti/U/Ti layers has already been described among others in a previous paper(1). The targets there were obtained by high vacuum evaporations of 1 to 3 $mg \cdot cm^{-2}$ of U onto rolled backing foils of 1 to 7 $mg \cdot cm^{-2}$ of Ti. After this, high vacuum evaporations followed to produce protective Ti films of about 0.05 $mg \cdot cm^{-2}$ in thickness on the surfaces of the freshly evaporated U layers. Thick backings were used at that time, because the Ti foils were needed as target supports and, in some experimental runs, also as windows in a gas jet system.

For further experiments sensitive substances as U, Pr, and some isotopically enriched isotopes of Nd and Sm had to be prepared as targets in thicknesses of a few $mg \cdot cm^{-2}$. In these cases it was necessary to keep the amount of shielding materials as small as possible. Therefore a new preparation procedure was developed. It starts with the mechanical and chemical cleaning of thick solid foils or pieces of metallic target material. If enriched isotopes of lanthanide elements are available in their oxide form only, they have to be reduced to the metal in a combined high vacuum reduction and distillation process before they are rolled out to suitable foils. The metal foils are mounted onto frames. They are cleaned further by glow discharge treatments in a high vacuum evaporation apparatus. Thereafter protective metals are evaporated onto both sides of the foils. As a result, sandwich layers are obtained which can be rolled down to a desired target thickness. Only well-known target preparation techniques are used in the whole procedure.

2. PREPARATION PROCEDURES

a) Preliminary Treatments of Target Substances

Pieces or thick foils of sensitive metals are sometimes delivered with oxide layers. They often prevent the materials from being oxidized completely. The oxide layers have to be removed prior to evaporations of shielding materials onto metal

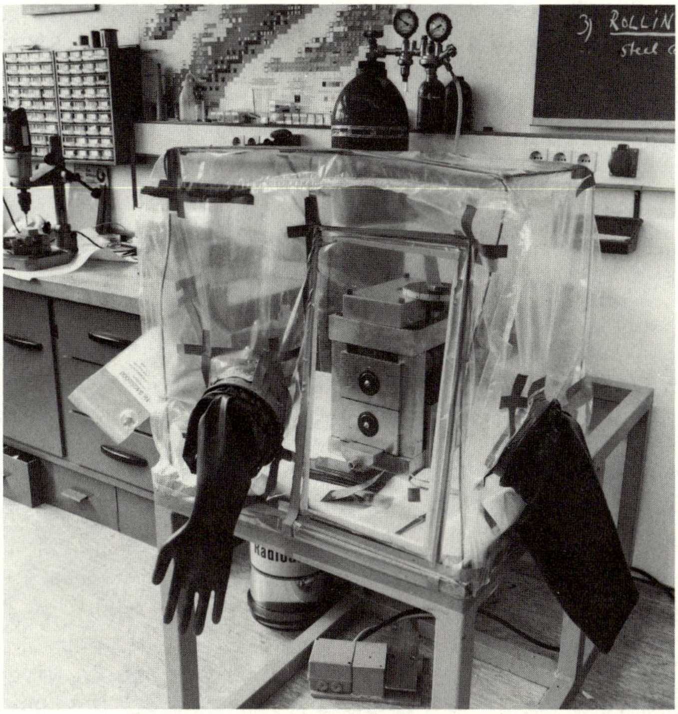

Figure 1. Rolling machine with glove-bag.

surfaces. If the desired target substances are available only as oxide powders, then they first have to be reduced to the metals. Thereafter they can be rolled out to a required size and thickness for subsequent treatments. Three different examples are given:

α) U

Foils of about 21 mg·cm^{-2} of depleted natural uranium (0.3-0.4% ^{235}U) were bought from NUKEM, Hanau, Germany. The surfaces of most foils had thin oxide layers which were dissolved in a bath containing cold concentrated nitric acid. After this the foils were rinsed twice with water, rinsed with isopropanol, and dried. They were mounted to frames and as soon as possible brought into a high vacuum evaporator for further treatments.

β) Pr, ^{142}Nd, ^{144}Nd, ^{146}Nd, and ^{148}Sm

The following procedures were performed in an improvised glove-bag made of plastic foil built around a rolling machine (see Figure 1). The glove-bag was flushed with Ar to maintain an atmosphere with an oxygen content of less than 0.5%. - The Pr foil was available in a thickness of 17 $mg \cdot cm^{-2}$ being stored in a sealed ampoule. The ampoule was opened and the foil was rolled down to 12.2 $mg \cdot cm^{-2}$. The isotopes ^{142}Nd, ^{144}Nd, ^{146}Nd, and ^{148}Sm were delivered in form of irregular pieces of 50 to 100 mg stored in paraffin-oil. Nevertheless, they had thin oxide layers which were removed mechanically by means of a scalpel. The pieces were rolled out to foils of a thickness of 7 to 15 $mg \cdot cm^{-2}$. All foils were packed into plastic boxes to be transferred into a high vacuum evaporator.

γ) ^{143}Nd

Isotopically enriched ^{143}Nd was obtained in its oxide form, so it had to be converted to the metal. This was accomplished in a combined reduction and distillation process using an electron beam gun evaporator received from G. Sletten, Niels Bohr Institute, Risø, Denmark. The reduction-distillation procedure and the special apparatus which was used have been described in detail by L. Westgaard and S. Bjørnholm(2).

The small evaporation device basically consists of a Ta crucible, 15 mm high and 4 mm o.d., having a 10-mm-deep drill-hole of 2 mm i.d. It is mounted onto a 30-mm-long Ta wire of 1 mm diameter. A positive voltage of about 1000 V is applied during a bombardment with electrons emitted from a heated helical Ta wire positioned around the crucible. The oxide reacts with a reducing agent added to the crucible to form the metallic lanthanide element which is evaporated and collected onto a cooled Ta condenser. The process is carried out under high vacuum conditions.

Thirty (30) mg of $^{143}Nd_2O_3$ powder were reduced in that apparatus by means of 48 mg of freshly powdered Hf which is about twice the stochiometric amount according to the equation

$$2\ ^{143}Nd_2O_3 + 3\ Hf \rightarrow 4\ Nd\uparrow + 3\ HfO_2.$$

Isolated metallic ^{143}Nd was rolled out to a 13.5-$mg \cdot cm^{-2}$-thick foil, using the above mentioned glove-bag. The foil was transferred in a plastic box to a high vacuum evaporator.

b) Evaporation of Protective Metal Layers

The rolled metal foils were positioned in a high vacuum evaporation unit type 770 from Veeco, Plainville, N. Y. It is equipped with a 6"-diffusion-pump. A glove-bag with an attached pass-through-chamber made of plastic foils was adapted to a Pyrex bell jar, 30" high and 18" diameter. Thus sensitive metals could be handled in an inert gas before they were covered with protective layers. The metallic foils were thoroughly cleaned by means of an intensive glow discharge, using 1.5-kV-Ar-ions at currents of up to 30 mA. A pressure of about 5×10^{-2} mbar was maintained in the apparatus. After this, protective layers of Ti or Ni were evaporated under high vacuum conditions from a 6-kVA-electron-gun type ESV-6 from Leybold-Heraeus, Hanau, Germany. The deposition rate was kept at about 5×10^{-4} $mg \cdot cm^{-2} \cdot s^{-1}$. Layers of 0.06 to 0.2 $mg \cdot cm^{-2}$ were evaporated onto both sides of the samples.

The obtained sandwich layers were ready to be stored for a long time or to be rolled down to the required target thickness.

c) Rolling of Sandwich Targets

A small machine is available to roll evaporated sandwich layers to a final target thickness. It is from Heiss, Dorfen, Germany, and it can be seen in Figure 1. The glove-bag which was described above was also used in the rolling of Ni/U/Ni samples

Figure 2. Rolling machine, steel sandwich, and some tools.

Table 1. Composition of Ni/U/Ni Sandwich Targets

TYPE	ORIGINAL TARGET MATERIAL		EVAPORATED LAYERS ON BOTH SIDES		ROLLED TARGET SANDWICHES		
Nr.	ISOTOPE	THICKNESS $mg \cdot cm^{-2}$	ELEMENT	THICKNESSES $mg \cdot cm^{-2}$ (x2)	TARGET $mg \cdot cm^{-2}$	PROTECTIVE LAYERS $mg \cdot cm^{-2}$ (x2)	%weight+)
1	^{238}U	21.0	Ni	0.07	0.8	0.003	0.7
2	"	"	"	"	1.0	0.004	"
3	"	"	"	"	2.0	0.007	"
4	"	"	"	"	5.3	0.020	"
5	"	"	"	"	7.1	0.025	"
6	"	"	"	"	8.2	0.030	"
7	"	"	"	"	10.7	0.035	"
8	"	"	"	"	13.3	0.045	"
9	"	"	"	"	15.3	0.050	"
10	"	20.6	"	0.06	11.0x)	0.030	0.6

+) Protective layers of both sides given as % in weight of the sandwich layers.

x) Series of 30 targets of 2.5x7.5 cm within a tolerance of 11.0±0.2 $mg \cdot cm^{-2}$.

to avoid radioactive contaminations of the laboratory. The glove-bag can easily be changed if required. Protected lanthanide elements were rolled down in the apparatus without using the gloves for reasons of simplification, but the prepared targets were stored under Ar or in an evacuated desiccator to be on the safe side. Figure 2 shows a detail of the apparatus. The machine has two driven rolls, 1200 mm long and 80 mm in diameter. The velocity is fixed to 20 rpm. All samples are rolled there between steel plates starting mostly with sheets of 0.5 mm thickness.

3. COMPOSITION OF TARGETS

Examples of Ni/U/Ni sandwich targets which were prepared for heavy ion bombardments at the UNILAC of GSI are listed in Table 1. Originally the U foils had a thickness of about 21 $mg \cdot cm^{-2}$. They received protective layers of 0.07 or 0.06 $mg \cdot cm^{-2}$ on each side and were stored in that form to be rolled down to a required target thickness just before an experiment. U targets varying in thicknesses from 0.8 to 15.3 $mg \cdot cm^{-2}$ were thus prepared having protective Ni layers of 0.003 to 0.50 $mg \cdot cm^{-2}$ on each side; in weight corresponding to 0.7% - in one case 0.6% - of the sandwich stack. It has to be pointed out, however, that Ni layers below 0.02 $mg \cdot cm^{-2}$ have only little shielding efficiency. Targets covered in this way must be used soon. In previous experiments U foils were also sandwiched in Ti layers, because an element with

Table 2. Composition of Sandwich Targets of Pr, ^{142}Nd, ^{143}Nd, ^{144}Nd, ^{146}Nd, and ^{148}Sm Being Covered with Layers of Ti or Ni.

TYPE	ORIGINAL TARGET MATERIAL		EVAPORATED LAYERS ON BOTH SIDES		ROLLED TARGET SANDWICHES		
Nr.	ISOTOPE	THICKNESS $mg\cdot cm^{-2}$	ELEMENT	THICKNESSES $mg\cdot cm^{-2}(x2)$	TARGET $mg\cdot cm^{-2}$	PROTECTIVE LAYERS $mg\cdot cm^{-2}(x2)$	%weight+)
1	Pr	12.2	Ti	0.16	3.5	0.05	2.6
2	^{142}Nd	7.5	Ni	0.1	"	0.05	2.7
3	"	9.7	Ti	0.1	"	0.04	2.1
4	"	6.7	"	0.1	"	0.05	3.0
5	^{143}Nd	13.5x)	"	0.2	"	0.05	3.0
6	^{144}Nd	12.0	"	0.15	"	0.04	2.5
7	^{146}Nd	14.9	"	0.16	"	0.04	2.1
8	"	14.2	"	0.15	"	0.04	2.1
9	^{148}Sm	12.0	"	0.2	"	0.06	3.3

+) Protective layers of both sides given as % in weight of the sandwich layers.

x) Obtained after reduction of oxide.

a small atomic number and a high melting point was required. These foil stacks could be stored for some months only, whereas Ni/U/Ni layers of a comparable thickness have already survived for more than a year. Ni apparently builds a very dense surface when evaporated onto U and it can easily be rolled together with U.

Target materials of the lanthanide elements, namely foils of Pr and of some isotopically enriched Nd and Sm isotopes, were obtained from preliminary treatments in thicknesses ranging from 6.5 to 14.9 $mg\cdot cm^{-2}$ (see Table 2). They were coated with Ti or Ni films of 0.1 to 0.2 $mg\cdot cm^{-2}$ on each side. All target sandwiches had to be prepared with a thickness of 3.5 $mg\cdot cm^{-2}$, thus protective layers of Ti or Ni of 2.1 to 3.3% in weight were received as can be seen from Table 2. Pr which is extremely sensitive to oxidation was destroyed already after some weeks in spite of its Ti shielding. Covering Pr with Ni will therefore be used in a future procedure. Isotopes of Nd and Sm protected with films of Ti could be stored unchanged in a desiccator for some months.

4. APPLICATIONS

Targets of 0.8 and 1.0 $mg\cdot cm^{-2}$ of U being protected by Ni films were used both in experiments on high energy δ-rays and on positron spectroscopy. The bombarding energies were 4.7 MeV/u for

^{208}Pb and 5.9 MeV/u for ^{238}U at intensities of 1.0 or 0.5 pnA, respectively(3). Different U targets ranging in thickness from 5.3 to 15.3 mg·cm^{-2} (See Table 1, types 4-9) were prepared to measure stopping powers in bombardments with ^{238}U ions of energies of 5.9, 8.5, and 10.0 MeV/u using a surface barrier-detector calibrated by a time-of-flight system(4). A series of 30 U targets of 2.5x7.5 cm were rolled out within a tolerance of 11.0 ± 0.2 mg·cm^{-2} (Table 1, type 10). The foils were mounted to a rotating wheel to distribute the beam intensity over a large target area in order to protect the targets from being destroyed at higher particle densities. Bombardments with ^{238}U ions of 7.6 to 8.3 MeV/u at currents of up to 150 pnA were performed with these targets in the search for superheavy elements with half-lives down to 10^{-3} s(5).

Pr, ^{142}Nd, ^{143}Nd, ^{144}Nd, ^{146}Nd, and ^{148}Sm were prepared in a target thickness of 3.5 mg·cm^{-2} protected with thin Ti or Ni layers (see Table 2). The targets were used for α-decay studies on very neutron deficient Tl and Pb isotopes employing an on-line mass-separation of evaporation residues from fusion reactions with ^{48}Ti beams of 4.9 to 6.6 MeV/u at current densities of up to 30 pnA(6). The beam was focused to 8 mm diameter. If the target temperature was kept below 800°C life-times of several hours were achieved with one target.

5. CONCLUDING REMARKS

The coating of oxygen sensitive substances such as U and a few lanthanide elements with evaporated metal films of Ni or Ti is an effective procedure to protect them from being destroyed rapidly. The resulting sandwich layers can be stored for a long time. They can as well be rolled down to a required target thickness without the use of inert gas systems.

The preparation procedure can be applied in general to all rollable target sandwich systems. The covering layers of the rolled sandwiches must be dense enough to build protective surface layers.

REFERENCES

(1) H. Folger and J. Klemm, in Proc. 6th Ann. Conf. of the INTDS, University of California, LBL-Report-7950 (1978) p. 69.
(2) L. Westgaard and S. Bjornholm, Nucl. Instr. and Meth. 42 (1966) p. 77.
(3) C. Kozhuharov, Private communication, GSI, Darmstadt, 1979.
(4) W. Brüchle, H. Folger, and J. V. Kratz, GSI-Annual Report 1979, in preparation.

(5) H. Gaggeler, E. Schimpf, W. Weber, and G. Wirth, GSI-Annual Report 1979, in preparation.
(6) U. J. Schrewe, P. Tidemand-Petersson, G. W. Gowdy, R. Kirchner, O. Klepper, A. Płochocki, W. Reisdorf, E. Roeckl, J. L. Wood, J. Zylicz, R. Fass, and D. Schardt, Physics Letters B, to be published.

PAST AND PRESENT TARGET MAKING

ACTIVITIES IN LABORATORIES IN THE UNITED KINGDOM

K. M. Glover

Actinide Orders
A. E. R. E. Harwell
Oxfordshire, OX 11 ORA
United Kingdom

Targets of separated stable and actinide nuclides were a priority requirement in the early days of the nuclear industry for the measurement of nuclear constants needed in reactor design and operation, fuel handling, storage and reprocessing.

1. A. E. R. E. HARWELL

(1) Stable Isotope Targets (Chemistry Division)

A large electromagnetic separator for the production of highly enriched stable isotopes commenced production on 8.2.50 and continued to operate until 14.11.67[1,2].

In 1953 the first separated stable isotope targets were produced. Early targets[3] were thin ($< 100\ \mu g/cm^2$) and made by vacuum evaporation onto backing materials such as Au, Cu and Ni. Early targets were also mainly of light elements, e.g., Li, Mg, Si, Cu and S. Electroplating[4] was used for the preparation of elemental targets of metals over a wide range of thicknesses. The technique was used for both backed and self-supporting foils deposited over precisely defined areas. Taking appropriate precautions a high degree of uniformity could be obtained and any unused and expensive isotope recovered. However, the method was not quantitative and accurate assessment of the quantity deposited posed problems. For light elements weighing was not sufficiently accurate and was unsuitable for reactive ones. The usual procedure was to plate to exhaustion and assay any residual isotope recovered from the plating cell. Choice of backing material presented problems, highly polished substrates resulted in poor adhesion and were therefore only used for the deposition of thick self-supporting foils. For other depositions the surface was electropolished or etched prior to deposition to ensure good adhesion. To ensure uniformity of deposit, the anode had to be flat, parallel to and as far removed from the cathode as possible, preferably larger than the cathode and either the anode or cathode should be rotated during deposition. Lacomit was used to define the deposition area. By 1960 cold rolling was introduced and where applicable became the preferred technique for the production of metallic foils.

Many of the techniques presently in use date back to the nineteen fifties. Vacuum evaporation is carried out using commercial evaporation units suitably modified to accommodate small isotopic samples. Except in situations where uniformity has been the prime consideration, efficiency has been maximized for expensive scarce isotopes by keeping the distance of the filament from the target to a minimum, usually 2.5 cms. Resistance and electron gun heating have been used. Self-supporting targets prepared over the years include, e.g., Sn, Li, Cu, Ag, Si, Ge, C, Fe, Ni, Ta.

Carbon backings, thickness in the range ten to several hundred $\mu g/cm^2$ have been evaporated onto a variety of substrates for the preparation of self-supporting foils using carbon arc.

Self-supporting targets of reactive elements such as Li, Ca, Ba, Sr, etc., in metallic form have been shipped in vacuum to avoid oxidation. Over the years techniques used(5) have included die compaction for powders, painting and sedimentation for refracting materials, molecular plating for rare earth oxides. Some of these methods proved less reliable than others.

Currently both backed and unbacked targets are made, preferably using one of the following techniques of proven reliability:

a) Vacuum evaporation (both by resistance heating and electron beam heating)
b) Cold rolling
c) Electroplating.

Efforts have been concentrated on improving the quality and extending the limits of these techniques rather than upon further fundamental development of new techniques. The present and future program include the use of hot rolling for hard metals such as molybdenum. The use of an argon arc button producer to prepare alloy beads for subsequent hot rolling, specifically produces alloy foils of varying composition for Mössbauer studies and structural damage measurements. Iron based alloys with 0.2 atom percent Sn-119 have been made with the elements Al, Sn, As, Cr, Cu, Ge, Au, Ir, Os, Pt, Rh, Si, Ti, W, V, and Zn in the range 1.5 atom percent. A similar series of nickel based alloys containing 0.2 atom percent Fe-57 have been made with the elements Al, Co, Cs, Ga, Ge, Mn, Si, I, and V.

Ion implantation is used for the preparation of standard layers of deuterium on a variety of backing foils and for the preparation of Rutherford back-scattering standards of Bi, Au, and Al into Si. Boron and lithium layers, thicknesses in the range 10-100 $\mu g/cm^2$ on a variety of substrates are also prepared by ion implantation.

(2) Actinide Targets (Chemistry Division)

The preparation of actinide targets for nuclear data measurements began at Harwell in 1950. The major requirement then was for uniform well-characterized uranium and plutonium deposits of known isotopic content, 1 mg/cm^2 thickness, for fission measurements. Painting(6) was selected as being potentially the most suitable technique to use to build up uniform adherent actinide deposits of this thickness. Initially to achieve maximum uniformity two different approaches were used. Uranium deposits were prepared by painting over the entire surface of a mirror finish platinum substrate. The required deposit thickness was achieved by the application of many coats, then the appropriate deposit area was covered with a thick layer of black wax and the entire foil immersed in 8M HNO_3 until the exposed uranium had dissolved. The foil was then washed with distilled water and dried. Finally the black wax was dissolved in CCl_4 leaving a uniform adherent layer with no edge effects. This method had the disadvantage that much heating and pumping were required to ensure complete removal of the CCl_4, which had a deleterious effect on counting gases in use at that time, also the method was not suitable for plutonium.

The first 99% enriched Pu-239 deposits were prepared by painting a platinum strip to a thickness of 1 mg/cm^2 and punching out the required area which was then soldered onto an inactive backing. The plutonium was assayed initially by weighing and finally on the individual sources by low geometry alpha counting(7). Agreement between the calculated Pu-239 per source calculated from the weighing and the alpha assay was within 1% indicating that good uniformity had been achieved. These sources were of very high quality, but the method was not readily adaptable to other deposit dimensions, was cumbersome for glove box operation and presented a decontamination problem, therefore all subsequent deposits were prepared using masks, taking great care to minimize edge effects.

During the nineteen fifties many actinide deposits of Pu-239, U-233 and U-235 were prepared by painting, including double-sided deposits on both platinum and aluminum backings of thickness up to 1 mg/cm^2, active areas up to 80 cm^2, and backing thickness from .010" platinum to 0.001" aluminum and 0.0005" nickel.

Back to back foils of Pu-239 and U-235 were prepared without cross-contamination. Uniformity of deposits was measured by autoradiograph and thickness variations across these deposits were found to be within ± 2%.

Fission foils were assayed by weighing, absolute alpha counting(7) and destructive analysis. Agreement between the methods was found to be within the error limits then specified, usually 1-2%. To reduce problems from inactive contaminants, special paint brushes were sometimes used consisting of glass fibers araldited into capillary tubing.

Fission chambers for flux monitoring, consisting of nests of concentric nickel cylinders were coated with Pu-239, 99% enriched, and U-235, 93% enriched, by dipping the nests in a lacquer solution containing a minimum of cellulose. The U-235 fission chambers were coated to 1 mg/cm^2 thickness 0.3 gm total uranium. Single cylinder fission chambers were coated again by painting. During this time other deposition methods were also tried. Uranium deposits 1 mg/cm^2 thickness were prepared by electrodeposition with limited success. The method was not quantitative, the deposits tended to be less uniform than painted deposits and reproducible conditions were difficult to achieve. Similar problems were encountered in the electrodeposition of plutonium.

During 1952 the availability of alpha spectrometry equipment using a gridded ion chamber with a multichannel pulse height analyzer resulted in a requirement for infinitely thin weightless actinide deposits and efforts were concentrated on development of techniques to meet the need.

During 1951 tetraethylene glycol (TEG)(8) was first used as a polymerizing agent in the preparation of thin uniform deposits for alpha spectrometry and has continued to be used to this day, for the preparation of quantitative assay sources. Resolutions of better than 20 KeV FWHM can be obtained using this technique.

Electrospraying(9) was tried, but with limited success. Vacuum evaporation was also tried and showed promise.

By 1958 resolutions of 20 KeV FWHM were attainable with signal generator pulses fed through electronics associated with a gridded ion chamber into a multichannel analyzer. Twenty-five (25) KeV FWHM was possible[10] with sources in the gridded ion chamber. This with the advent in the early nineteen sixties of surface barrier detectors led to increased emphasis on the need for a reliable technique for the preparation of thin uniform deposits of a whole range of actinide nuclides for alpha spectrometry and other nuclear data applications requiring thin deposits uncontaminated by other nuclides.

Hermes, the actinide separator for plutonium isotopes commenced operation in 1957, so that separated plutonium isotopes were becoming available by the late nineteen fifties. Hermes continued in operation until 1964.

By 1959[11] the basic design of a resistance heating vacuum evaporation apparatus suitable for evaporating actinides had evolved. That design with modifications and improvements continues in use to the present day. The basic philosophy is of a simple system with interchangeable evaporation chambers which ensure that by having a separate evaporation chamber for each nuclide, deposits of separated actinide nuclides can be prepared without any possibility of cross-contamination.

Deposits of actinide nuclides giving resolutions of a few KeV can reliably be prepared by this technique. Sources for alpha spectrometry give resolutions of less than 11 KeV FWHM peak height and Cf-252 fission fragment energy sources conforming to the Schmidt & Pleasonton criteria are prepared on both thin and rigid backings. Deposits (e.g., 50 μg Ra-226, U-232 and Th-228) have been prepared on thin carbon and mylar backings and on large single crystals.

During the nineteen sixties, methods were sought to maximize efficiency and still reliably produce thin uniform deposits of rare and expensive heavy actinides. Electrospraying[12] was pursued further and uranium deposits up to 1 mg/cm^2 were produced, also thin deposits of Am-241, but the latter had a relatively poor resolution (17.5 KeV FWHM) compared to vacuum evaporated deposits. It proved impossible to achieve reliably reproducible operating conditions compatible with the required flexibility, so the use of the electrospraying for source preparation was discontinued.

Molecular plating was also used to prepare large area deposits on aluminum backing foils, but again it proved difficult to establish reliably reproducible operating conditions. The deposits tended also to have poor adherence to the substrate.

Currently, vacuum evaporation is used to prepare all deposits for nuclear data measurements from weightless up to 200 μg/cm^2 on both rigid and thin backings. Painting is used to prepare deposit thicknesses in the range 200 μg to several mg/cm^2.

Present activities include the preparation of both thick and thin deposits of the entire range of available actinide nuclides on both rigid and thin, e.g., 100 μg/cm^2 thickness nickel substrates. There is a particular demand for the latter type of source. The levels of activity requested frequently cause problems in manufacture, transport and shelf life. Considerable ingenuity may be required to resolve these difficulties. At the other end of the scale there is an increasing demand for large area deposits of many actinides up to 1 mg/cm^2 thickness for fission and other nuclear data measurements.

It is interesting to note that carefully controlled targets and sources prepared by painting or vacuum evaporation more than ten years ago are still safely used today, e.g., 30 mCi unsealed Pu-238 source used to irradiate biological samples, fission foils and counter calibration standards, 500 μg/cm^2 thickness uranium deposits on 0.002 mm thickness aluminum substrates prepared for cyclotron irradiation.

(3) Non-Isotopic Targets (Nuclear Physics Division)[13]

Target making started during the nineteen fifties with the commissioning of the 5 MeV Van de Graaff accelerator. A wide range of non-isotopic targets were produced by vacuum evaporation using both resistance heating and electron gun. For example, targets of thickness up to 200 μg/cm^2 were produced from the following elements: Fe, V, Bi, Pb, W, Ag, Zn, Ir and Cu. Carbon foils 5 μg/cm^2 thickness were prepared by carbon arc for use as stripper foils in the Van de Graaff accelerator.

Carbon foils were also used as substrates in the preparation of all non-isotopic targets.

Present day activities additionally include the preparation of lithium fluoride deposits, thicknesses in the range 50-100 μg/cm^2 on silver substrates. Thin layers of tin, chromium, hafnium and silica thicknesses of the order of 300 angstroms are prepared by electron gun heating for ion implantation applications. There is also a requirement for self-supporting aluminum foils of high uniformity with thicknesses in the range 10 μg to 4 mg/cm^2.

(4) Some Additional History

In 1951 a symposium on thin films sponsored by Chemistry Division, Harwell, was held at Buckland House (ancient manor house). The objective was to bring together users and producers and to discuss types of films and methods of preparation. Topics discussed included:

1. criteria for judging films, e.g., quantitative assay, stability of deposit, uniformity, area definitions, accuracy of isotopic analysis, and quality of substrate with preference for thin deposits

2. substrate stability of deposits, adhesive and cohesive forces

3. summary and discussion of (already numerous) methods for thin film preparation including thermal decomposition by van Arkel, precipitation reactions, evaporation of solutions and suspensions, use of tetraethylene glycol and pyridine, electrode position, cataphoresis, monolayer deposition, crystal growth, thermal evaporation, sputtering, electromagnetic separation, rolling, heating, dipping, spraying and painting. (Question: How much have we really progressed in 30 years?)

Further reports from the preparative group of the Chemistry Division are listed under references (30 to 36).

2. ATOMIC ENERGY ESTABLISHMENT, WINFRITH

(1) The assessment of reactor assembly performance by precise experimental methods calls for samples for irradiation and samples for post irradiation counting of high isotopic and chemical purity in the form of thin uniform deposits (of finite area) to avoid and minimize self-shielding effects[14]. At Winfrith in the nineteen sixties, electrodeposition techniques were found to be very suitable for the preparation of such deposits. The deposition conditions of a wide variety of isotopes of interest, e.g., Pa-231, U-234, U-235, U-236, Np-237, Pu-241, Pu-242 and Am-243 were investigated. Deposition conditions were studied using perspex or polythene cells with cylindrical platinum anodes and end windows of either <3 microns thick stainless steel or gold, the former having a transmission of approximately 60% for 5 MeV alphas, whereas the latter have only approximately 30% transmission. Seals were made with VITON. Electrolytes such as formic/formate, nitric acid or ammonium chloride were used and

it was found that the addition of low concentrations of oxalate initially were beneficial in terms of source quality. Current densities of 0.125 mA/cm^2 and up were used requiring external EMF's of between 7 and 15 volts. The pH was monitored by means of a micro-sampling device. The rate of deposition was monitored by a zinc sulphide screen in contact with the end window and connected to a chart recorder via a photomultiplier and ancillary electronics. Quantitative primary calibration foils for Zebra were produced by this method and the accuracy of the work confirmed by low geometry alpha counting.

The first successful quantitative deposits of Pa-231 and Pa-233 ever reported(15) were prepared at Winfrith on stainless steel backings from any one of the following electrolytes:

a) 0.2M ammonium fluoride
b) 0.015M ammonium fluoride, 0.25M ammonium chloride
c) 0.02M, oxalic acid, 0.06M nitric acid.

Usually 50 μg of natural uranium were used as a carrier and co-deposit. Polythene cells were used with neoprene gaskets. Freshly cut gaskets were conditioned before use by boiling in concentrated hydrochloric acid with subsequent water leaching. Daughter products were removed from the Pa-231 and the Pa-233 was separated from its parent Np-237. Both were rigorously purified before use. The quantitative nature of the deposition method was determined for Pa-231 by α-counting and for Pa-233 by γ-counting. Using this technique deposits were prepared on highly polished stainless steel source mounts from solutions containing HF. The stainless steel was not flamed before or after deposition.

(2) Large area uranium deposits up to approximately 200 μg/cm^2 thickness on aluminum backings were successfully prepared from solvent solutions for neutron and heavy ion irradiation, where the low capture cross-section of Al is of great advantage. This would not have been possible from aqueous media. An in-depth study was made of the mechanism of the process and the amount of material deposited with respect to such parameters as time, current, integrated current voltage and amount of uranium in the cell. The effect of certain additives was examined, and the deposit was analyzed for nitrate and uranium. The solubility of the deposited material was investigated qualitatively in several solvents. Additionally, the effect of removing dissolved oxygen from the electrolyte was studied. The results of the analysis of the deposited compound indicated that the uranium is deposited as a solvated and hydrated form of uranium trioxide.

(3) Electrospraying was used from time to time.

(4) Current work is directed towards coating large numbers of fission chambers and a non-flammable lacquer has been developed for glove box work with actinides and for which a patent is pending.

3. ATOMIC WEAPONS ESTABLISHMENT, ALDERMASTON (NUCLEAR PHYSICS DIVISION)

A Special Techniques group was set up at AWRE during 1956 to provide support to the main accelerator facilities. These included the first Tandem Van de Graaff constructed, a 3 MeV Van de Graaff, the HVEC C. N. 5.5 MeV accelerator as well as other smaller facilities. The activities covered by the Special Techniques group included, target making, vacuum techniques, accelerator tubes, nuclear counters and detectors, etc. Considerable demands were made on the target group for the preparation of a wide range of targets. As the expertise developed, requests for targets were received from universities without commercial facilities and a commercial organization was established to deal with these requests(16). Targets were dispatched world-wide to destinations which included Australia and South Africa. A special vacuum dessicator packaging was developed to permit the transport of delicate stripped and ready mounted targets. Targets were also prepared to customer request outside the range listed if possible.

(1) Preparation of Thin O-17 and O-18 Targets(17)

Targets of O-17 and O-18 were prepared from enriched WO_3. Two cleaned filaments of tungsten were outgassed by heating for four hours at a pressure of $< 1 \times 10^{-6}$ torr, then enriched D_2O was carefully admitted and the WO_3 produced was deposited on the chamber walls. On completion of the reaction, the WO_3 was carefully brushed off the walls and subsequently evaporated onto self-supporting carbon backings (5-20 $\mu g/cm^2$) from a 1/8" diameter platinum cylinder, and spot welded onto a tungsten filament. Thin films up to 500 $\mu g/cm^2$ were produced by this method.

(2) Self-supporting C-13 foils(18) were prepared by thermal decomposition of methyl iodide enriched in C-13 and collection of the C-13 enriched carbon on a 0.0005" nickel substrate. The nickel substrate was cleaned and outgassed by passing an electric current through it until it reached bright red heat at a pressure

approximately 1.4 x 10^{-8} torr. When the vacuum stabilized the enriched methyl iodide was introduced into the system until a pressure of 40 torr was reached. After approximately two minutes the methyl iodide remaining was recondensed for recovery, the carbon coated nickel foil was removed from the reaction chamber and the nickel stripped using 10% H_2SO_4, 20% HNO_3 and 70% H_2O. Finally, the self-supporting carbon foils were mounted on suitable target frames.

(3) Film Thickness Measurements(19)

The following techniques were used at various times for measuring the thickness of thin self-supporting foils:

Weighing, using a semi-microbalance
α-Particle thickness monitor
Modulated beam photometer
Densitometer
Crystal thickness monitor
Dynamic balance
Vapor beam ratemeter.

(4) Alpha and Fission Counting of Thin Foils of Fissile Material(20)

Alpha foils of less than 0.1 mg/cm^2 thickness were counted to an accuracy approximately 1% at 2π geometry. Fission foils thickness in the range 0.1-1.0 mg/cm^2 could be assayed to an accuracy of 0.2-1.0% depending on foil thickness and the application of foil thickness corrections. The relationship between substrate surface, deposition technique, deposit thickness and counting errors was uncertain.

The work at AWRE ran down in the late 1960's and the target activity was licensed to Vacuum Instruments and Products of Newbury, Berkshire by the U. K. A. E. A. This firm discontinued production after a few years due to falling demand.

References (21-29).

Dr. R. Middleton, the scientist in charge of the Tandem facility left AWRE in 1963 to become Professor of Nuclear Physics at Pennsylvania State University, Philadelphia and with AWRE support created a Target Laboratory which subsequently provided a Commercial Target Service in the U. S. A. similar to that previously provided by AWRE.

Mr. Muggleton is now at Canberra University preparing targets for the accelerator physicists there.

(4) UNIVERSITY OF OXFORD

Target making started about 1958. Early targets of B and C were prepared by vacuum evaporation using an electron gun.

Present methods used include vacuum evaporation using both resistance heating and electron gun; rolling, electroplating and sputtering. The future program includes direct target preparation using ion beam implantation.

(5) UNIVERSITY OF MANCHESTER

Target making began in 1967. Stable and active targets have been prepared over the years including U-235, Cs-137, C-14 and Tritium. Techniques used include electroplating and vacuum evaporation. Rare earth oxides are reduced to metal and subsequently rolled.

(6) UNIVERSITY OF LIVERPOOL

Some targets were prepared during 1960-65, but from 1965 onwards all targets for the University's experimental programs were prepared in its own laboratory.

During the early days most targets were of light elements (e.g., Si, Mg) on gold backings. Later self-supporting thin targets (200-300 $\mu g/cm^2$) were prepared using vacuum evaporation with resistance heating.

Methods now in use include:

vacuum evaporation (by resistance heating and electron gun)
electroplating
rolling
some sputtering.

(7) THE RADIOCHEMICAL CENTRE, AMERSHAM

During 1951/52 a laboratory was set up at The Radiochemical Centre, Amersham for the preparation of actinide targets. Small sources were prepared by two methods, painting and using TEG as a spreading agent. A 2 mg Ra-226 source was prepared on an aluminum substrate approximately 4 x 1 cm. In 1953 in an effort to improve the quality and uniformity, electrodeposition was used. The method chosen was based on the oxidation of Pu-IV to Pu-VI. Pu-VI remained stable in a slightly acidic media used. During electrodeposition the Pu-VI was reduced and Pu-IV deposited on the cathode. This technique was found to be successful for deposits up to 1 mg/cm^2 thickness. Americium and Curium deposits were also prepared by electrodeposition using methods available in the literature, e.g., ammonium oxalate and oxalic acid. It was found that as the thickness of the deposit increased the efficiency of deposition decreased and it seemed that this method was only applicable to relatively low specific activity nuclides.

Uniform Po-210 alpha sources, strengths 1-10 mCi, were prepared by electrodeposition from 0.5M HNO_3.

From 1960 vacuum evaporation was used to prepare alpha sources.

The following present day activities of TRCL may be of interest.

A diverse range of radiation sources for industrial and laboratory applications have been prepared at the Centre for many years. Radioisotope sources of low energy gamma and X-radiation are finding increasing application in many diverse fields of industry, medicine and research. Significant development in instrumental techniques is extending and improving these applications, particularly in the fields of thickness gauging and X-ray fluorescence.

Source materials commonly used are Am-241, Pu-238 and Cm-244, which when incorporated in a ceramic matrix enable useful sources to be prepared with emitted radiation in the energy region 10-60 keV. High energy β sources, e.g., Sr-90 are used for transmission gauging, e.g., in controlling the amount of tobacco in cigarettes or air in ice cream.

Gamma sources, e.g., Cs-137 find application in transmission gauging where the material to be gauged is too thick or too dense for the use of β sources, e.g., to control the speed of material being fed into a hopper. In packaging, this technique finds extensive use in the production of detergents, soaps, pharmaceuticals or the filling of drink cans.

Alpha foil as used in smoke detector sources is prepared by a rolling technique in which the radioisotope, as an insoluble, non-volatile compound is uniformly dispersed in a matrix of pure gold or silver sintered at a high temperature. This is contained between a backing of silver 0.2 mm thick and a front covering of gold 0.003 mm thick. The high integrity containment of the radioactive material allows the foil to be subdivided into small pieces for use in various industrial applications.

For laboratory applications β, γ, and X-ray reference sources are prepared and calibrated.

8. 20TH CENTURY ELECTRONICS/PREPARATIVE CHEMISTRY GROUP, HARWELL

In 1956 20th Century Electronics Division was licensed by UKAEA to produce fission chambers and foils for nuclear data measurements. Initially 20th Century set up a hot laboratory to handle natural uranium, depleted uranium, enriched uranium and thorium deposits.

In 1959 in collaboration with the Preparative Chemistry Group at Harwell, U-233, Np-237, Pu-239, Pu-240 and Pu-241 deposits were prepared and a new hot laboratory was commissioned at 20th Century Electronics to handle these nuclides.

Evaporation and painting techniques were used. To coat the internal surface of a fission chamber 0.040" diameter, the evaporation filament consisted of a wire on which the active nuclide had been deposited and located on the axis of the tube to be coated. This novel technique regrettably did not prove too successful and painting was resorted to. Evaporation methods were tried including crucibles heated by RF, but the efficiency of this technique was too low. A painting technique using a felt pad to coat the inner surfaces of fission chambers was tried, but excessive loss of activity in the pad and contamination of the outer surface of the cylinder led to the abandonment of this method. Finally, painting, using a brush was adopted as standard.

Present day activities are restricted to deposition of low specific activity nuclides, e.g., natural uranium and thorium.

CONTRIBUTORS

Mr. G. A. Barnett, AEE, Winfrith
Mr. F. A. Howe, AWRE, Aldermaston
Mr. D. Boreham, AERE, Harwell
Mr. M. Watts, AERE, Harwell
Mr. L. J. Bint, AERE, Harwell
Mr. M. Nobes, AERE, Harwell
Mr. J. B. Reynolds, University of Liverpool

Mr. T. L. Morgan, University of Manchester
Mr. D. Dwight, TRCL
Mr. R. Alan, 20th Century Electronics Ltd.
Mr. K. L. Wilkinson, AERE, Harwell

REFERENCES

(1) M. L. Smith, Electromagnetically Enriched Isotopes and Mass Spectrometry, 13-16 September 1955, Harwell.
(2) D. H. Randall and M. L. Smith, Nature, June 11, 1955.
(3) J. B. Reynolds and D. Boreham, Experience of Vacuum Evaporated Targets, Proceedings of the Seminar on the Preparation and Standardization of Isotopic Targets and Foils, AERE R 5097, Harwell (1965).
(4) F. A. Burford, J. H. Freeman, and J. B. Reynolds, AERE R 5097.
(5) K. M. Glover, F. J. G. Rogers, and T. A. Tuplin, Nucl. Instr. and Meth. 102 (1972) pp. 443-450.
(6) K. M. Glover and P. Borrell, J. Nuclear Energy, Vol. 1 (1955) pp. 214-217.
(7) G. R. Hall and K. M. Glover, Precision α-Particle Counting, Nature 173 (1954) p. 991.
(8) R. Hurst and K. M. Glover, Note on the Use of Tetraethylene Glycol (TEG) as a Spreading Agent in the Preparation of Uniform Plutonium Sources, AERE C/M 100.
(9) D. J. Carswell and J. Milsted, J. Nuclear Energy, Vol. 4 (1957) pp. 51-54.
(10) K. M. Glover, IAEA Symposium on Standardization of Radionuclides, SM 79/39 (1966).
(11) N. Jackson, Vacuum Sublimation Apparatus for the Preparation of Thin Sources of α-Active Materials, J. Sci. Instr., Vol. 37, May 1960, p. 169.
(12) P. S. Robinson, Nucl. Instr. and Meth. 40 (1966) p. 136.
(13) M. Nobes, J. Sci. Instr., Vol. 38, October 1961.
(14) G. A. Barnett, J. Crosby, and D. J. Ferrett, AERE R 5097, p. 28.
(15) G. Smith and G. A. Barnett, J. Inorg. and Nucl. Chem., Vol. 27, (1965).
(16) Technical Information Bulletin No. 2 (undated) Special Techniques Group AWRE.
(17) A. H. F. Muggleton, Preparation of Thin ^{17}O and ^{18}O Targets, AERE R 5097.
(18) A. H. F. Muggleton, A Method for the Preparation of Thin Self-supporting ^{13}C foils, AERE R 5097.
(19) F. A. Howe, A Critical Survey of Methods for Determination of Film Thickness Applicable to Nuclear Targets, AERE R 5097.
(20) P. H. White, Alpha and Fission Counting of Thin Foils of Fissile Material, AERE R 5097.

(21) A. H. F. Muggleton and F. A. Howe, A Crystal Oscillator Film Thickness Monitor, N. I. M., Vol. 28(2) (1964) pp. 242/244.
(22) A. H. F. Muggleton and F. A. Howe, The Preparation of Thin Film Boron Films, N. I. M., Vol. B(2), pp. 211/214.
(23) AWRE Report No. 32/67, The Production of Nuclear Targets by Vacuum Evaporation, May 1967, 40 pages.
(24) A. H. F. Muggleton and F. A. Howe, The Preparation of Thin ^{18}O Targets, N. I. M. 12 (1961) pp. 192-194.
(25) G. T. Arnison, The Production of Thin Self-supporting Nickel Foils using an Electroplating Method, N. I. M. Vol. 53(2) (1967).
(26) A. H. F. Muggleton and Parsons, N. I. M. Vol. 27(3) (1964) p. 357.
(27) G. T. Arnison, A Technique for Producing Deuterated Polyethylene Targets, N. I. M. Vol. 40(2) (1966) p. 359.
(28) G. T. Armson and R. J. Gilmour, The Preparation of Thin Self-supporting Films using a Heated Substrate, N. I. M. 45 (1966) p. 178.
(29) D. P. Gregory, The Preparation of Thin Uniform Deposits of Uranium and Plutonium by Electrodeposition, etc., AWRE Report No. 0 29/56 also AWRE Report No. 0 5/57.
(30) E. Furby and K. L. Wilkinson, Vacuum Evaporation of Thick Films of U - Operated in a glove box. Some experiments with Pu, (2.8 mg/cm^2), AERE C/R 2441 (1958).
(31) D. J. Dwight, The Preparation of Pu-239 Alpha Sources by Electrodeposition Source Thickness up to 1 mg/cm^2 claimed, RCC/R75 (TRCL now) (1959).
(32) F. Hudswell, The Preparation of Thin Films and Coatings. Discussion of Available Techniques, AERE C/R395 (1949)
(33) J. Biram, Adherent Evaporated Films concerned with Substrate Pretreatment, AERE EL/M 86 (1954).
(34) R. H. Dawson and K. L. Wilkinson, Plutonium: Evaporation Tests, Ionization Potential and Electron Emission. Project Concerned with Crucible Materials and Deposits for Ion Source Development, AERE GP/R 1906 (1956).
(35) M. G. Schomberg, Preparation of Beta Sources on Very Thin Backing Materials by Vacuum Evaporation - Preparation of intense Au-199 Sources, AERE M479 (1959).
(36) J. Stephen, An Electrolytic Method of Preparing Mössbauer Sources and Absorbers - Sources Prepared by Introducing Co-57 into Natural Iron or Cobalt, for Diamagnetic Metals, Copper, Gold and Platinum are used, AERE R4308 (1963).

PREPARATION OF NUCLEAR TARGETS AT THE INSTITUTE OF ATOMIC ENERGY

Sun Shu-hua, Su Shih-chun, Chen Qing-wang,
Guan Sheu-ren, and Xu Guo-ji

Institute of Atomic Energy
Academia Sinica
Peking, China

ABSTRACT

In this paper, the development of nuclear targets for nuclear reaction experiments in the Institute of Atomic Energy, Peking, China, is described. The techniques of vacuum evaporation, electroplating and isotopic material produced by electromagnetic isotope separators are discussed. The number of different kinds of targets produced are about thirty and they are in the thickness range of a few $\mu g/cm^2$ to several mg/cm^2. About one-hundred isotopes of twenty-six elements have been produced in this Institute.

1. INTRODUCTION

To meet the requirements in nuclear physics research at our Institute, a laboratory for the preparation of nuclear targets was established in 1974. Several people who had some experience in preparing targets were gathered. Various techniques have been developed and many apparatus have been set up and a large number of targets have been prepared and supplied to nuclear reaction research and some other fields since then.

Vacuum evaporation and electroplating are the two main methods used in the preparation of targets in this laboratory. In the early state, we concentrated only in the techniques to prepare targets with natural light elements. Later on, we gradually turned our attention to the preparation of the isotopic pure and some medium weight elements targets.

The thickness of the prepared targets are from a few μg/cm^2 to several mg/cm^2.

At the same time, electromagnetic separators have been used to produce enriched isotopes in our Institute, and some isotopic targets have since been made.

2. TECHNIQUES USED IN THE PREPARATION OF NUCLEAR TARGETS

2.1 Vacuum Evaporation

Vacuum evaporation is the most common method in the preparation of nuclear targets. In our laboratory, three evaporators and other apparatus were used in this method. The vacuum in the evaporator was kept to better than 2 x 10^{-5} torr. Both resistance heating and electron bombardment were used as the evaporation heat source.

Table 1. The Targets Prepared by Resistance Heating Method

Element	Type	Substrate	Parting Agent	Thickness (μg/cm^2)	Diam. (mm)
Li	Metal, LiF	C		200	18
Be	Metal	S. S.	Soap	20- 50	12
O	Mo_2O_3	C			18
F	CaF_2	C		200	18
Na	Metal	C			18
Mg	MgO	C		200	25, 18
	Metal	S. S.	KCl, Al	300	
Al	Metal	S. S.	Soap, CsI	20- 500	18, 25
Si	Monocrystal	S. S.	Soap, CsI	20- 50	18
P	Red Phosphorus	C, Au		40	18
Cl	$BaCl_2$	C		100	18
K	KCl	C		100	18
Ca	CaF_2	C		300	25
	Metal	S. S.		1000	
V	Metal	S. S.	Soap, CsI	25- 60	18
Se	Natural	C		100	18
Rb	RbCl	C		100	18
Ag	Metal	S. S.	Soap, CsI	50-1000	18
I	CsI	C		100	18
Cs	CsI	C		100	18
Ba	$BaCl_2$	C		100	18
Au	Metal	S. S.	Soap, CsI	50-1000	18, 25

C-carbon film, S. S.-self-supporting target.

2.1.1 The Targets Fabricated by Resistance Heating Method are listed in Table 1. For example:

(1) Magnesium (Mg)

The targets of isotopic magnesium (^{24}Mg, ^{25}Mg, ^{26}Mg) were prepared by reduction evaporation method(1). In this case, magnesium-oxide was used as the starting material and zirconium powder as the reducing agent. Four pieces of carbon films Ø 18 mm were suspended about 6 cm above a tubular type tantalum crucible with a total charge of 6 mg isotopic MgO in the crucible. Forty (40) $\mu g/cm^2$ metallic Mg can be deposited upon carbon film in this way. Under this condition, the collection of isotope was about eleven percent.

Self-supporting magnesium targets were prepared in the following way: clean thin Al foil was stuck on a flat glass slide, then a thin layer of KCl was evaporated on Al, and finally metallic Mg was deposited upon KCl by direct evaporation. The deposited Mg film can be easily stripped off by floating in distilled water. The thickness of the self-supporting target thus prepared can be as low as 300 $\mu g/cm^2$.

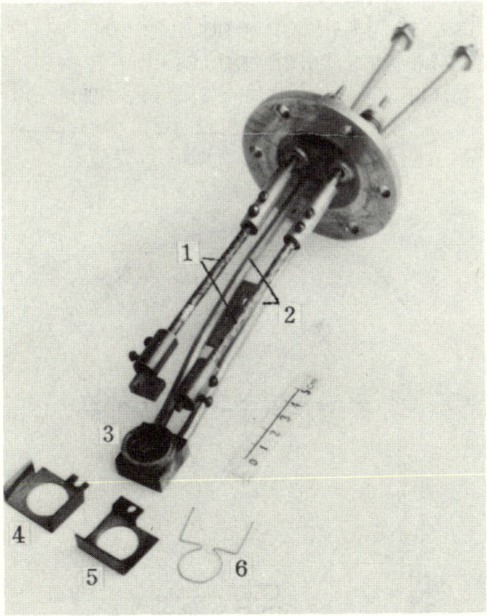

Figure 1. The structure of electron gun. 1) The two holders, 2) water cooling tubes for crucible, 3) crucible, 4) the top of focusing box, 5) the bottom of focusing box, 6) filament.

(2) Silicon (Si)[2]

Flexible and durable self-supporting Si film of 20-50 $\mu g/cm^2$ can be produced by direct evaporating Si upon glass slides with predeposited KCl as separating agent, but the distance between crucible and glass plate should be larger than 30 cm, otherwise the evaporated Si film will be brittle and suffer from cracks.

(3) Vanadium (V)

Vanadium targets were made by direct evaporation of V from a boat shape tungsten crucible. Since vanadium and tungsten would turn to alloy at high temperature, several evaporating runs were required at one evaporating process with small quantities of target material, and in each run different crucibles were used.

2.1.2 Electron Bombardment Method. Another method of source heating employed was electron bombardment, which was often reserved for the more refractory materials.

The electron bombardment apparatus consists of three parts: electron gun, power supply and vacuum system. The electron gun is shown in Figure 1. It includes a filament with its two holders, an electrostatic focusing box, and a water cooling crucible. The filament is a circle of Ø 0.5 mm tungsten wire. Focusing electrode made from tantalum foil 1 mm thick and in the form of a box with openings Ø 19 mm at its top and bottom side, encloses the filament. The crucible was machined from a whole piece of oxygen free copper in the form of a 25 x 25 x 10 mm^3 block with a

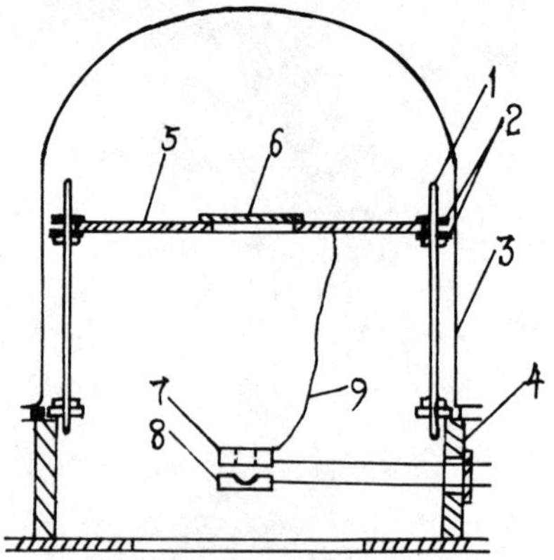

Figure 2. Schematic drawing of evaporation set. 1) Prop., 2) insulator, 3) glass jar, 4) transition ring, 5) metal plate, 6) glass slide, 7) focusing box, 8) crucible, 9) lead wire.

Ø 20 mm x 1.8 mm sample hole drilled at its center, and installed at 10-20 mm beneath the focusing electrode.

Both the current for heating filament and negative potential of 0 to 6 kV are drawn from the holders of filament. The crucible is connected to the ground and serves as the anode as shown in Figure 2.

The electrons emitted from the filament pass through the bottom hole of the focusing box and are thus focused to bombard the material in the crucible. The focusing spot is 3-4 mm in diameter.

Targets of boron(3), vanadium and cobalt were prepared by this apparatus.

Raw material of boron is a black powder. A 5 ton oil press is used to press the boron powder into small pellets of 20 mm diameter and 3-4 mm thickness before putting it into crucibles to prevent the scattering of the powder sample during bombardment.

The preparation of cobalt target is alike A. Rose's method(4).

2.1.3 Carbon film (C). Thin carbon films were fabricated by arc-discharge evaporation. Self-supporting targets could be produced by the use of a suitable separating agent. Films of good mechanical strength can be obtained by using soap solution as separating agent; films of very good flatness and low oxygen contaminant can be obtained by using CsI as separating agent. In both cases the thickness of the C-films could be as low as 10 μg/cm^2.

2.2 Electroplating

This technique has the advantage of economy in comparison with methods mentioned above but generally it produces films of poor uniformity and purity if not under good conditions(5). In our Laboratory, we designed and constructed an electroplating cell for making small quantities of material. Self-supporting targets of Co, ^{64}Zn and ^{63}Cu were prepared by this method.

(1) ^{64}Zn

The area of target is 3.8 cm^2 and the thickness is 1.7 mg/cm^2. The prescription of the electroplating solution is as follows:

ZnO	70 mg/10 ml
0.1N H_2SO_4	8 ml/10 ml
$(NH_4)_2SO_4$	500 mg/10 ml
$KAl(SO_4)_2$	50 mg/10 ml

Glycerine	1/20 ml
pH	3.8

The operating conditions are as follows:

Current density	3-5 mA/cm^2
Voltage	2-3 V
Anode	Pt (spired type plate of metal Pt wire with a 1 mm diameter)
Cathode	Al (the foil thickness is 0.2 mm)
Temperature	Room
Distance between the two electrodes	3 cm

(2) ^{63}Cu

Target area is 3.8 cm^2 and the thickness is 0.5-2 mg/cm^2. The prescription of the electroplating solution is as follows:

CuO	35 mg
In H_2SO_4	45 ml
15N HNO_3	0.5 ml
pH	0.6-0.8

The operating conditions are as follows:

Current density	6-10 mA/cm^2
Voltage	2-3 V
Anode	Pt
Cathode	Stainless Steel
Temperature	10-20°C
Distance between the two electrodes	3 cm

2.3 Tritium-Titanium Target

Tritium-Titanium target is a target in which T was absorbed by a thin titanium film predeposited on some substrate and serves as the nuclear reaction target. Here titanium acted only as a carrier of tritium, therefore its quality mainly depends on the nuclear properties of tritium.

Fourteen (14) MeV neutron can be produced by the reaction of $T(d,n)^4He$. This neutron source has a suitable energy and a well-known energy spectrum. Therefore it is very useful in nuclear reaction studies, activation analysis, neutron cross-section measurements and radiation damage studies.

Figure 3. Photograph of the separators.

Titanium was deposited on a molybdenum substrate by vacuum evaporation at a pressure of about 2×10^{-5} torr. Typically the thickness of the titanium film is 0.2-3 mg/cm^2. The Mo substrate, coated with titanium film, was then placed in the vacuum chamber and tritium gas was introduced into the chamber and heated to about 400°C by H.F. heater and tritium was rapidly absorbed by titanium. The pressure of the tritium is 150-200 mm Hg in the chamber. The targets obtained with this method are used in the accelerators and have a half-life of 3 mA/hr for 14 mm diameter targets with 200 keV, 100 μA deuteron beam and 10^9 n/sec neutron yield, and 10 mA/hr for 50 mm diameter targets with 360 keV, 3.45 mA deuteron and 3.6×10^{11} n/sec neutron yield(6). Targets of various sizes were made in this way.

3. MATERIALS WITH ENRICHED ISOTOPES

The isotopes produced by the large-scale electromagnetic isotope separators are still considered as the main source of stable isotopes. Laboratory isotope separators have the advantage that thin targets can be prepared to high isotopic purity by deposition from laboratory isotope separators. In recent days a large number of laboratory isotope separators have been installed in various laboratories all over the world.

Two large production machines(7) (i.e. calutrons) were constructed in 1965 at our Institute. They are the standard 180 focusing mass separator with linear shims, the mean radius of curvature is 90 cm and 170 cm, respectively. Figure 3 is a photograph of the two separators. The ion source generally used is the high intensity source with lateral extraction. They contain an oven for heating the charge and increasing its vapor pressure and a discharge chamber for stripping the electrons from the element. During the operation in recent years ten to several tens milliampere collecting current could be obtained.

Besides these, a Scandinavian type low intensity isotope separator(8) with high resolution was constructed in an earlier stage. The analyzer magnet is a 90° device, with a 160 cm radius. Nielsen source and Freemen source were designed and constructed for different requirements. They give several hundred μA and one mA ion current, respectively.

Nuclear targets are prepared by placing the substrate at the collector position in the laboratory type separator. Typically such as silver isotopes of 100 $\mu g/cm^2$ thick can be deposited on aluminum foil of 5 μm thickness by the sputtering method and with isotopic purity ≥99%. Lithium and potassium isotope thin targets of small quantities are prepared by direct collection at full energy (40 keV).

During the past 15 years we separated about one hundred isotopes of twenty-six elements in our Institute. Only part of them have since been used in target preparation.

REFERENCES

(1) A. Szalay, Nucl. Instr. and Meth. 49 (1967) 355.
(2) A. M. Sandorfl and D. R. Kilius, Nucl. Instr. and Meth. 136 (1976) 395.
(3) A. H. Muggleton and F. A. Howe, Nucl. Instr. and Meth. 13 (1961) 211.
(4) A. Rose, Nucl. Instr. and Meth. 35 (1965) 165.
(5) J. M. Heagney, Nucl. Instr. and Meth. 102 (1972) 451.
(6) Hau Bin-kan, Private communication, Lanchou Modern Physics Institute.
(7) Koch, J., Electromagnetic Isotope Separators and Applications of Company, 1958.
(8) G. Andersson, Ark. Phys. 12 (1957) 331.

UNIVERSITE LOUIS PASTEUR - STRASBOURG (FRANCE)

Centre de Recherches Nucléaires
Group PNIN: M. Weishaar
Group PNPA: A. Meens
Group Spectrométrie Nucleaire: M. A. Saettel
23, rue du Loess, 67037 Strasbourg Cedex

In the 1950's, when the Centre de Recherches Nucléaires was located in the hospital complex, there was one novice target maker (with one evaporator) who met most of the modest target needs of the period.

Soon after the installation in 1960 and rapid growth of the Centre in its present location in Strasbourg-Cronenbourg, four target laboratories have been created, but one was closed in 1976. They evolved from their early circumstances of no equipment and no experience to the more or less completely equipped target laboratories of today, which were, and still are, attached to three group leaders. The three technicians (two with chemical and one with electro-mechanical background) are called on to prepare a large variety of targets for approximately 80 physicists at the four Van de Graaff accelerators in Cronenbourg and sometimes in other laboratories.

The techniques used include evaporation (joule heating, electron gun), electro-deposition, rolling, cracking, etc.... With largely standard techniques, the target makers can satisfy the needs with only a few exceptions.

As for the future, this is of course difficult to predict. With the high energy accelerators under construction or projected, it may be that target requirements will be different. Thus in the future, as at present, new techniques will be studied and eventually implemented as the need for them arises.

THE MODULAR TARGET TRANSPORTATION AND STORAGE FACILITY, VAC

J. H. Bjerregaard, P. Knudsen, and G. Sletten

The Niels Bohr Institute
University of Copenhagen
Denmark

ABSTRACT

Extensive use of external accelerator facilities has emphasized the need for a safe and handy target transportation container. The present paper describes a portable vacuum container 45 mm high and 95 mm in diameter with capacity for twelve 20 x 25 mm target frames. During transportation the container can maintain a vacuum better than .1 torr for 24 hours and below 1 torr for a week. The container is adaptable to a central pumping station where long time storage in the laboratory takes place at 10^{-6} torr. Each one of the 10 containers connected to this station can be removed or pumped out without breaking the vacuum of the others.

INTRODUCTION

During the last 5 years an increasing amount of the experimental physics work at our Institute has been involving external accelerator facilities. Whether these experiments take place at the Brookhaven Tandem Facility or at GSI in Darmstadt, they usually imply transportation of several delicate target foils.

Apart from destruction by mechanical shocks, some of these targets perish under normal atmospheric conditions by the influence of O_2, CO_2, H_2O and possibly other components. Especially after exposure to accelerator beams the corrosive effects of atmosphere become serious for most materials.

Therefore, there have been several good reasons for the development of a target transportation container. In the design

Figure 1. Transport container without targets. A 20 x 25 mm target frame with a ø = 15 mm window is shown.

we have demanded the following 3 criteria to be fulfilled:

1) Mechanical stability for the targets during transport.
2) Exclusion of corrosive media to the 10^{-4} level for at least 24 hours.
3) Low weight and small dimensions, and with room for a dozen standard targets.

In the following we describe a container that fulfills these criteria, and which has been in regular use during the last 3 years.

A further development has made the container a module in a long term storage system which we for short call VAC.

Mechanical Properties of the Transport Container

Figure 1 shows the container together with the target frame it is designed for. The total height is 45 mm and the diameter of the cylindrical part 95 mm. The main body of the container is made of aluminum, and the lid is 6 mm thick glass. The vacuum seals are a 5 mm Edwards Speedivalve and a 75 mm diameter O-ring between the aluminum body and the glass lid. The lid is held firmly up to the O-ring by a screw cap. Including the Speedivalve the largest outer dimension of the container is 140 mm. The valve handle also extends 25 mm above the container lid. The total weight is 0.75 kg.

Inside the container targets are held in position by parallel slots. These slots overlap about 1 mm with the outer edges of the frames in each side. The slots are machined with sufficient tolerance to slide targets in with ease, but at the same time precise enough to keep the targets from rattling back and forth during transportation.

When the targets are loaded in the slots, these can be blocked by a sliding metal plate, one on each side of the frame. These slides can be fixed firmly by screws.

Vacuum Properties

The container can be connected to any pumping station by suitable adapters from the 5 mm Edwards Speedivalve. The main pumping station for the complete VAC system can maintain 10 containers at 10^{-6} torr for long time storage.

The pressure inside the container is determined by a leak of the order 10^{-7} torr · liter/sec. This means that a container which is disconnected from the central vacuum can maintain pressures below 10^{-1} torr for 24 hours. After a week the pressure will have increased to about 1 torr.

It is important for the performance of the isolated container that the residual atmosphere is dry. During tests a perforated vial containing about 1 gram of molecular sieves was introduced in the container.

With the 3 mm aluminum calcium silicate pellets the rise in pressure inside the isolated container was determined exclusively by the 10^{-7} torr · liter/sec leak.

It is believed that this leak can be diminished since there probably are only two main sources for the leak, namely porosity of the aluminum ingot and deficiencies of the O-ring seals.

The Long Term Storage Facility

The target transportation containers also enter as modules in

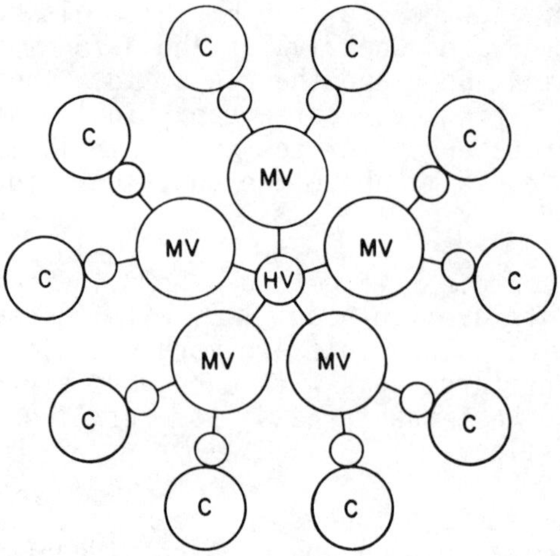

Figure 2. Layout of the long term storage facility VAC. HV means high vacuum, MV multivalve and C transport container.

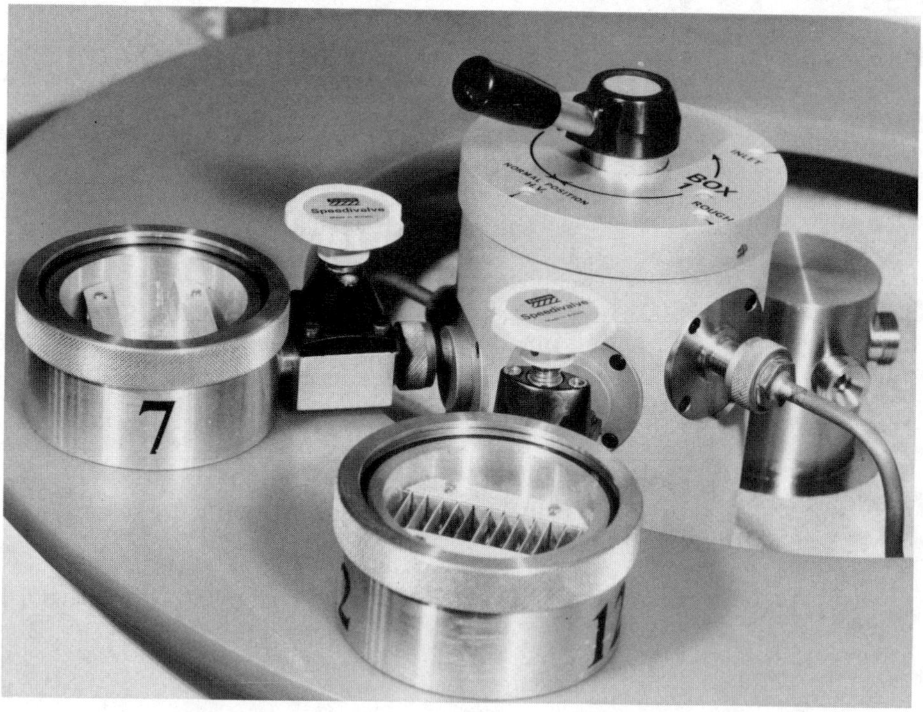

Figure 3. Detail of a pair of containers connected by the multivalve to VAC.

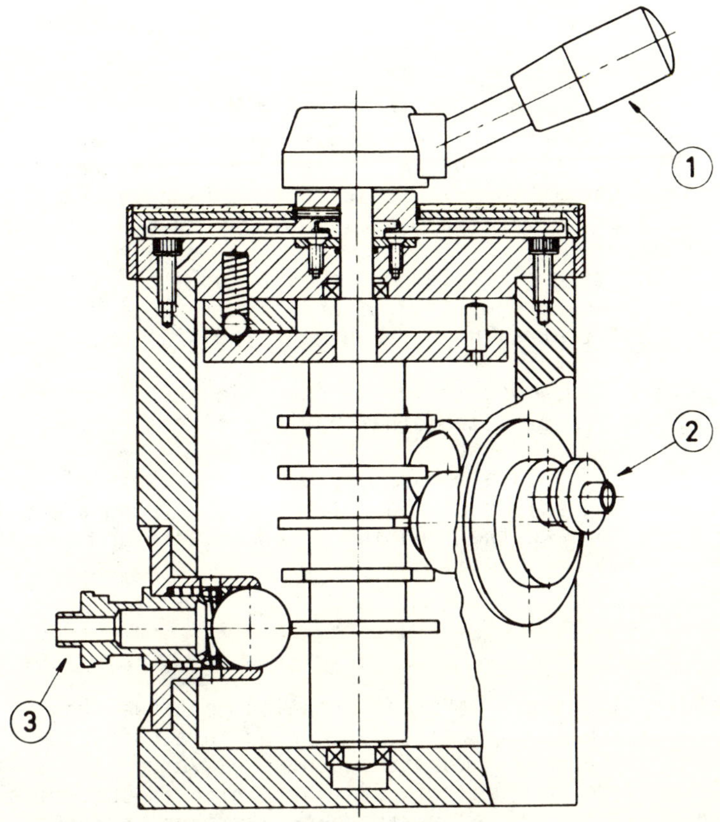

Figure 4. The multivalve for VAC.

1) Handle for changing mode.
2) Adapter for transport container.
3) High vacuum line.

a long term storage facility at our Institute. Of the 15 containers existing at the moment, 10 can be connected to a central pumping station where they are maintained at a vacuum of 10^{-6} torr by a turbomolecular pump. This system we call VAC, and Figure 2 shows the general layout of the containers and valves.

Figure 3 shows a part of VAC with two of the transport containers connected to it. As shown in Figures 2 and 3 these two containers are connected to the central vacuum by a common-valve, and altogether there are 5 such multivalves. As it will be described in the following it is possible to mount, pump and dismount each one of the 10 containers independently by means of this special valve.

The multivalve is shown in Figure 4 and serves a pair of containers A and B. In sequence the following modes are possible:

1) High vacuum pumping to both A and B.
2) After closure of the Speedivalve on B, B can be dismounted after a dry argon venting to its adapter. During this operation A is sealed off at high vacuum while the 8 other modules have active high vacuum pumping all the time.
3) Remounting of B, if necessary, roughing out of B.
4) High vacuum pumping of B and simultaneous reopening of A to the central high vacuum pumping station.

The valve is symmetric in its operation and permits the same operational sequence to the other partner of the pair.

VAC is built into a laboratory desk, and the whole ensemble of containers and valves can be rotated 180° around a vertical axis in order to make the mounting of containers and operation of the valves easier.

Conclusions

A small, light target transportation container has been designed. It is small enough to be carried in a briefcase and can accommodate 12 targets. When the container valve is closed, the pressure rises to 10^{-1} torr in 24 hours which is sufficient for most laboratory to laboratory travels. Further increase to about 1 torr takes one week.

As a general purpose target container these vacuum properties fulfill our demands. It is possible to travel with self-supporting rare earth targets like Nd and metallic Ca foils for at least a week with visually undetectable reduction of their quality. Some of these reactive targets require loading and opening of the container in an inert atmosphere glovebox.

The modular system VAC for long term storage permits physicists to have their own little target bank maintained at 10^{-6} torr ready for the upcoming experiment. Whether the experiment takes place next door or on another continent really does not matter any more.

THE RECOVERY OF METALLIC MERCURY

J. L. Gallant

Atomic Energy of Canada Limited
Chalk River Nuclear Laboratories
Chalk River, Ontario, Canada, K0J 1J0

ABSTRACT

Liquid mercury is recovered from its partial amalgam with copper by heating to the dissociation temperature and condensing the vapor on the walls of a reactor vessel. The technique has been used to recover small quantities of isotopically enriched mercury.

INTRODUCTION

Expensive quantities of liquid mercury 196, 198, 199, 200, 201, 202, and 204 targets are used in such experiments as the measurement of g-factor of isomeric state in the range of 1 ns to 1 ms. The mercury metal is prepared by heating the oxide in air(1). This paper is concerned with the preparation of the substrate and the recovery of the mercury from its amalgam with copper.

Preparation of the Substrate

The solubility of mercury in copper at 20°C is 2×10^{-3} weight %; this is a rather slow reaction. However, if the copper surface is reacted with nitric acid (cleaning the copper), the mercury will spread uniformly over the entire surface and form a bond which is difficult to break. Hence, the mercury must be contained in a small designated area of the copper surface. This is accomplished in the following way.

(a) (b) (c)

Figure 1 (a) Copper substrate with depression at center.
(b) Copper substrate chromium electroplate.
(c) Mercury target on chromium electroplated copper substrate.

A suitable copper target holder 0.8 mm thick is prepared with a depression at the center of 4 mm diameter and 1.5 mm deep (Figure 1a). This depression is masked with a tape. Chromium is electroplated on the remaining surface (Figure 1b) in a chromic acid bath.

The mercury (usually 50 mg) is dropped into the depression which contains a drop of dilute nitric acid (1:3). The acid is washed immediately with distilled water (Figure 1c).

The mercury is contained in the depression and will not amalgamate with the chromium which has a mercury solubility of $< 4 \times 10^{-7}$ weight % at 20°C.

Mercury Recovery

When a mercury target has deteriorated beyond its usefulness, the expensive isotope can be recovered in the following manner:

The copper frame is sheared leaving only the area holding the mercury. The mercury on copper is placed in a reactor vessel

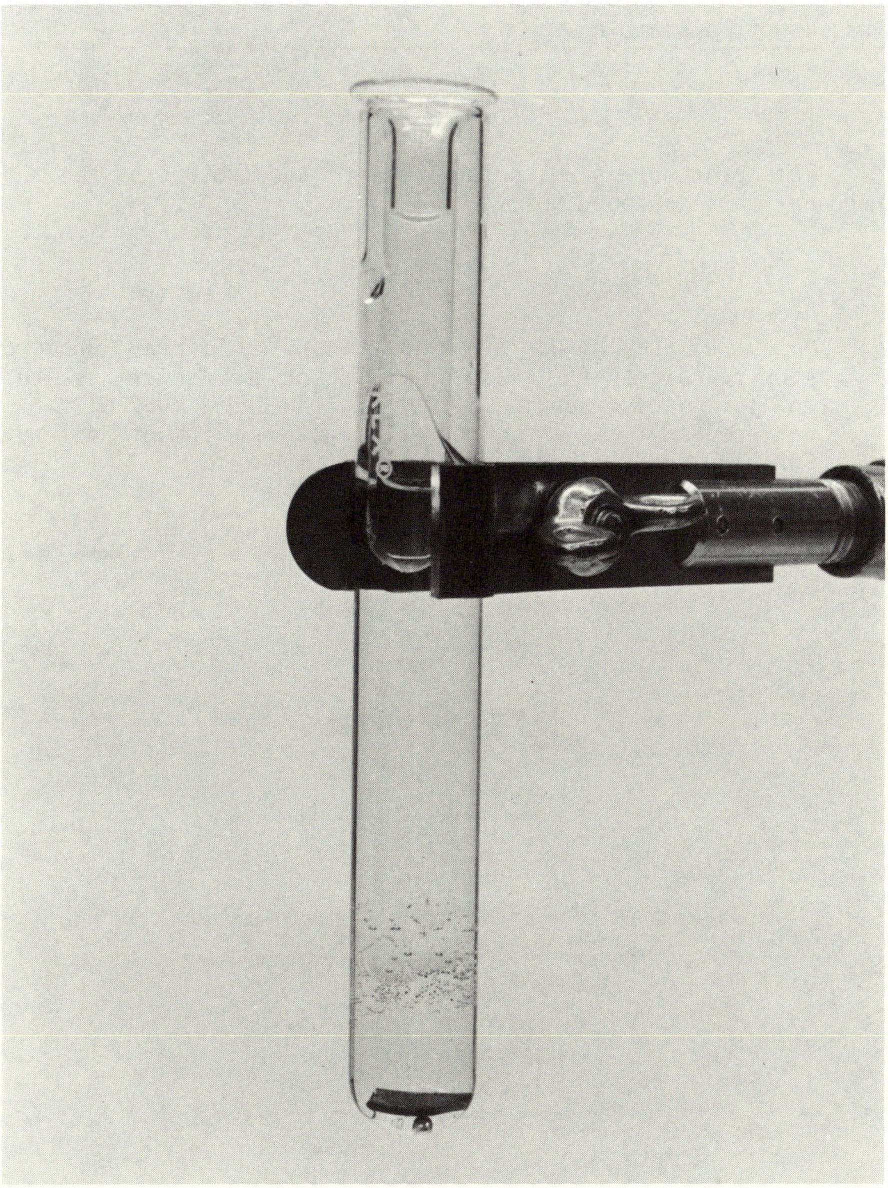

Figure 2. Mercury being recovered in reactor vessel.

(Figure 2) and heated until the dissociation takes place. The mercury condenses on the walls of the vessel. The copper metal is removed without disturbing the mercury, the vessel is filled with distilled water and the mercury scrubbed from the walls of the vessel with a spatula.

Conclusion

This procedure permits the reuse of the isotope as the recovery rate is above 90%.

REFERENCES

(1) J. L. Gallant, D. J. Yaraskavitch, and N. C. Bray, Preparation of targets for measuring g-factors of isomeric states and for fission studies of muonic ^{235}U and ^{238}U, World Conference of the Int. Nuclear Target Development Society, Garching, Sept. 1978.

NEW ENGLAND NUCLEAR CORPORATION

Steve Kendall, John L. Need, Robert MacKay, and Jozef Jaklovsky

New England Nuclear Corporation
North Billerica, MA 01862

New England Nuclear Corporation is the world's leading manufacturer of radioactive chemicals for research and radioactive pharmaceuticals for medical diagnosis. Additionally, it is a principal supplier of radioactive chemicals for clinical diagnosis, liquid scintillation counting chemicals and products used to calibrate diagnostic and research instrumentation.

Founded by two scientists in 1956, New England Nuclear is a private, non-governmental concern dedicated to helping researchers and clinicians obtain knowledge which will better our lives.

During its first ten years, NEN manufactured radioactive chemicals for use by researchers who otherwise would have to prepare these substances in their laboratories. The company grew steadily, largely because of the talent and dedication of people like Edward Shapiro, Ph.D., one of the founders and presently chairman of the board. The demand for NEN's chemicals was great, not only due to their consistent high quality, but also because the United States' increased its emphasis on medicine and technology during this period.

In 1966, the company recognized the growth potential of a second field: nuclear diagnostics. Consequently, NEN began to develop innovative radiopharmaceuticals for medical diagnosis. By 1978, the company was the leader in this field, primarily due to its ability to conceive, develop, produce and rapidly distribute these pharmaceuticals throughout the world.

Liquid scintillation counting chemicals, radioimmunoassay kits and reference sources -- NEN's three other product areas --

were introduced in the 1960s. While they represent a smaller proportion of the company's production than radioactive chemicals and radiopharmaceuticals, these product groups have expanded steadily. They are among the most complete available in the world.

NEN distributes its products worldwide, where forty countries constitute its routine users. The company is well-respected and, since its founding, has had sales growth of more than twenty percent each year. In fiscal year 1980, the growth rate was 24 percent.

One measure of NEN's success is the quantity and quality of its personnel. Between 1977 and 1980, for example, the number of employees grew from 900 to almost 1,500. Today, approximately 12 percent have advanced university degrees, including many persons who hold a Ph.D. in the sciences. Many of the chemists who were important to the company's early growth still work at NEN.

A second measure of NEN's success is its acceptance in 1979 of two IR-100 Awards, which are presented each year to the companies responsible for the United States' 100 most significant scientific developments. New England Nuclear was one of just a few organizations which won more than one award. The awards are the result of work by researchers within the company who continually interact with researchers and diagnosticians to better understand their needs and develop products to meet those needs.

Over its history, New England Nuclear has built facilities and resources which enable it to grow to meet various customer needs.

Company headquarters is in Boston, Massachusetts, which is also the location of NEN's radioactive chemicals for research manufacturing facility.

Thirty miles north of Boston, NEN owns more than 200 acres of land. This is the site of its Nuclear Medicine and Technology complex, which includes manufacturing facilities for radiopharmaceuticals, radioimmunoassay kits and reference sources.

Within its manufacturing facilities, NEN has the most modern equipment for research, production and quality control. In the radiopharmaceutical area, for example, the company has a significant advantage over other organizations because it owns three cyclotrons, which produce two of the isotopes needed most in medical diagnosis. A fourth cyclotron will begin production in mid-1980, and a linear accelerator will be completed in 1982. The linear accelerator will be the first such machine owned by a commercial organization, and it is expected to have at least eight times the production capacity of one cyclotron. Moreover,

Figure 1. NEN's first cyclotron with lid up for service. Ion source and target probes in foreground.

Figure 2. A vial containing radioactivity in an aqueous solution - one of the many ways NEN provides its products.

the accelerator will enable production of isotopes which are not feasible to produce by other means.

The challenge that was met in using these cyclotrons was the sequential development of several generations of targets suitable for the production of large quantities of radioisotopes. The current production rates in mCi/hr. are some 20-30 times as high as those we first obtained. The goal that is always ahead of us is the development of a high current, low effective power-density and easy to fabricate target. We have made modest progress.

Before products are shipped from NEN, they must pass rigorous quality control testing for both chemical and radiochemical purity. The consistent quality of the company's products is one of the major reasons for NEN's excellent reputation.

NEN's products are sold to its North American and European customers by representatives who are directly employed by the company. In other parts of the world, agents who are familiar with the needs of researchers and clinicians represent NEN.

Because New England Nuclear's customers often have specific requirements and many of the company's products have a short half-life, a highly efficient distribution network has been developed. NEN is able to ship its products all over the world while they remain useful, even in the case of radioisotopes such as thallium-201, whose half-life is just 73 hours. Approximately 7,000 shipments are made by the company each week

NEN manufactures over 1,200 different radioactive chemicals, including approximately ninety new products each year. They're used by nearly 40,000 of the world's biomedical researchers.

In addition to radioactive chemicals which are routinely available, NEN's custom synthesis service meets particular needs by radioactively labeling a wide range of chemicals which are not listed in the company catalogue.

NEN's radioactive chemicals are used for research in numerous areas, including cancer, heart disease and mental illness. The company manufactures radioactively labeled amino acids, steroids, carbohydrates, lipids, nucleic acids, catecholamines, drugs and numerous others. Currently, two areas of significant product growth are products for recombinant DNA and brain chemistry studies.

Radiopharmaceuticals account for the largest proportion of NEN's growth. The company's leadership in this business is largely due to superior manufacturing technology, innovative products and marketing programs, and the efforts of dedicated sales, distribution and customer service personnel.

Among NEN's products is thallium-201, for which NEN was the only manufacturer in the world as of early 1980. This revolutionary heart imaging agent can indicate if heart disease is present; if coronary bypass surgery is necessary; what areas will benefit from surgery; and, after surgery, how well the heart is functioning.

NEN's other radiopharmaceuticals include gallium-67 for diagnosing cancers of the lung and lymphatic system, and for detecting infection; the Technetium Tc-99m Generator, for brain imaging, and for the preparation of other imaging materials; OSTEOLITE™ and PYROLITE™, for bone imaging; GLUCOSCAN™, for kidney and brain imaging; and PULMOLITE™ and xenon-133, which assist in the detection of pulmonary embolism and other lung ailments.

Radioactive chemicals for clinical diagnosis, NEN's third major product group, includes radioimmunoassay kits, reagent paks and antisera. Their accuracy, reliability and economy make them popular throughout the world.

Products include tests useful in the diagnosis and management of prostrate cancer; pituitary-adrenal disorders; iron metabolism disorders; fetal status in high-risk pregnancies; hypothyroidism; plasma renin determination; and levels of the drugs digoxin (for heart conditions) and gentamicin and tobramycin (antibiotics).

HISTORICAL SUMMARY OF TARGET TECHNOLOGY AND THE INTERNATIONAL NUCLEAR TARGET DEVELOPMENT SOCIETY

E. H. Kobisk, President

International Nuclear Target Development Society

The incentive to organize the International Nuclear Target Development Society came as a natural consequence of many persons' activities associated with nuclear physics over a span of more than 40 years. With the advent of subatomic particle accelerators it was a necessity to establish a preparation function to obtain "targets" into which charged and neutral particles, γ-rays and x-rays, neutrons, and ions could be directed for the purpose of studying "particle"-nucleon interactions. Targets therefore are simply prescribed physical arrays of nuclei (usually of an isotope) designed to permit detection and measurement of specific interactive events at known energies.

In the early years of nuclear physics research (1940-1960), most targets were designed to permit accelerator beam transmission through a very thin array of nuclei with relatively few interactions taking place; however, some targets were very thick or were supported on a thick substrate (backing). These latter samples were designed to totally stop incident particles usually to produce transmutations of target nuclei or to observe backscattered product particles or electromagnetic radiations. Considering the wide variety of target forms necessary to perform meaningful research and the large number of elements (isotopes) to be studied, it is apparent that much technology in material science, metallurgy, chemistry, and physics was brought to bear on the problems of target preparation.

In general, target forms now include planar films from 200 Å thick to massive blocks of large geometric dimension (usually for high energy physics research), wires, rods, and other geometric shapes fabricated from metals, oxides, halogenides, intermetallics, cermets, and ceramics. Both stable and radioactive species of every element in the Periodic Table (natural and man-made) have been prepared in some target form.

With the advent in 1943 of isotope separations in large quantity (>10 mg) by mass spectrographic methods in machines known as calutrons, the function of target preparation immediately became more complex and far more costly. The value of separated isotopes vastly exceeded that of natural or mixed isotopic species, e.g., normal calcium can be purchased for $0.01/g while ^{48}Ca (in compound form) is presently priced at $376/mg or a factor of 37,600,000 more expensive. Not only is the cost problematic, but abundance and availability of isotopes are such that utmost precautions must be taken to conserve these materials and to maintain their chemical and nuclidic integrity. Thus the target maker is faced with the almost inordinate difficulty of having to adapt existing technology, known to produce satisfactory target samples from normal assay elements, to the preparation of samples using minute quantities of starting material--an adaptation which frequently fails and requires the development of new technology.

Almost every research installation throughout the world having an accelerator or a research reactor has personnel employed in the business of target preparation. Personnel assigned to this function have varied from Nobel laureates to auto mechanics, from persons with multi-PhD level scholastic backgrounds to those having high school level academic achievement. Not only is the educational level of target maker personnel varied, but the disciplines of their training broadly range from theoretical physics, through most of the natural sciences and engineering, to the less scientific areas such as business administration. It is apparent that discipline and scholastic achievement of target makers are not necessarily guideposts of success in this endeavor, but rather cleverness, innovativeness, and tenacity are more important; at least this certainly was the case in the early years of target preparation.

Since many target samples have similar characteristics and preparative methodology has shown signs of being definable, it has behooved target development personnel to share their unique methods so as not to waste valuable time and materials "re-inventing the wheel". To this end the United States Atomic Energy Commission (USAEC) and the Central Bureau of Nuclear Measurements (BCMN) at Geel, Belgium, began a liaison of target technology and distribution of isotopes in 1963. Dr. George Rogosa and Mr. James Garrett of the USAEC together with the author journeyed to Geel for the initial personal exchange of information. At that meeting, another target development person (Dr. McElroy) from Harwell, U. K., was present and contributed greatly to the success of the meeting. It should be noted that BCMN personnel and facilities at that time were relatively new, but the personnel were obviously very talented and their equipment and technology was impressive. The technology that had been developed at BCMN included such techniques as electrospraying and levitation melting and vaporization, techniques, at that time, totally unique to that installation.

At Harwell, U. K., highly sophisticated target methodology had and was being developed; these efforts were supported by two isotope separators similar to the original calutrons at Oak Ridge, Tennessee (USA). Under the direction of Dr. Harry Freeman, an extensive target facility had been established. The technologies developed at Geel and at Harwell were not redundant, but complimentary, a practice which could provide far more capability in target preparation than simple parallel achievement. However, for the most part availability of target samples from either installation was highly limited to the locale of the respective installations or to member nations of EURATOM. In both cases, however, occasional samples were prepared for researchers in other countries, particularly if the need or end use seemed to justify the expense.

Coincidentally, a six-man target preparation group had been formed and was active at the Oak Ridge National Laboratory (ORNL), USA. Initial efforts for a unified target preparation function were begun at ORNL about 1958 under the direction of Dr. Leon Love, director of calutron operations for isotope separation. The basic purpose of this activity was not simply to provide target samples to the ORNL accelerators, but to provide isotope targets to the world research community since the availability of isotopes from the ORNL calutrons was several orders of magnitude greater than supplies anywhere else in the world. It was the USAEC motive of cooperative distribution of isotopes to other countries that spawned the new target center. When I entered this organization in 1962, approximately 35 targets were being produced per year under the direction of Mr. B. J. Massey and the function had been relocated in the Isotopes Division of ORNL (including the calutrons) under the direction of Mr. John Gillette. This Division was responsible for separation of stable and radioactive isotopes, sources and target preparation, and related research and development. Because of health reasons I replaced Mr. Massey in 1962.

Upon revisiting installations at BCMN; Harwell; Amersham, U. K.; Saclay, France; Risø and Copenhagen, Denmark; and Stockholm, Sweden; in 1965, it was evident that considerable growth in the function of target preparation had transpired and that cooperation amongst these research centers and others throughout the world was needed. By this time, world recognition of the availability by purchase of isotopes and targets from ORNL was widespread (target production had increased to several hundreds of samples per year at ORNL). Similarly, target and materials standards preparations at BCMN had expanded significantly. Visitations of non-U. S. collaborators to Oak Ridge produced considerable camaraderie over the next several years and many new target groups became visible in these exchanges. Extensive target preparation groups had been established at the major USAEC sites of Lawrence Livermore Laboratory (LLL), Lawrence Berkeley Laboratory, Brookhaven National Laboratory, Los Alamos Scientific Laboratory, and at Argonne National Laboratory.

In 1963, a "Research Materials Meeting" was held at ORNL to initiate an exchange of information concerning isotope technology and the uses of isotopes; personnel from many USAEC Laboratories and from Chalk River (AECL), Canada, and several other non-U. S. countries attended this symposium. The information exchange included source fabrication, target fabrication, isotope separations, treatment of radioactive materials, and the function of material availability and sales from ORNL.

The success of information exchange at ORNL led to the formation of a Research Materials Committee under the auspices of Nuclear Physics and Chemistry, USAEC. Twice a year this committee, composed of appointed members of most AEC Laboratories (about 30 persons) convened for the purpose of information transmittal on new developments and on the fate and disposition of stable and radioactive isotopes and special materials useful in research. The success of this Committee in locating "lost" materials, in recognizing the needs of research and finding appropriate materials has continued over a span of 17 years and still is active in its functions. From this Committee has come guidance in isotope separation needs and even international exchange of materials which target preparations personnel have used extensively.

Continued liaison amongst target preparations experts incited the first effort to assemble interested persons on an international basis for the purpose of technology exchange and to identify problems associated with materials, techniques, shipping, and sample availability. In 1971, Oak Ridge National Laboratory and the USAEC co-sponsored a three-day meeting in Gatlinburg, Tennessee (U. S.) for target fabrications personnel and special isotope separations experts from 12 countries. The meeting was attended by about 90 persons. Excellent formal exchange of information was augmented by productive exchanges on a private or group discussion basis. The importance of the information interchange was evident and in the ensuing years a continuing annual assemblage of international target development personnel occurred according to the schedule noted in the table shown below.

Date	Place	Host(s)
May 1972	University of Montreal, Canada	E. H. Woodburn
October 1973	University of New York, Stonybrook, New York (USA)	W. Dan Riel
October 1974	Atomic Energy of Canada Limited, Chalk River, Canada	Joe Gallant Wayne Perry
September 1975	Argonne National Laboratory, Chicago, Illinois (USA)	Joe Thomas Frank Karasek
October 1976	Los Alamos Scientific Laboratory, Los Alamos, New Mexico (USA)	Judith Gursky John Povelites
October 1977	Lawrence Berkeley Laboratory, Berkeley, California (USA)	Gordon Steers
September 1978	University of Munich, Technical University Munich-Garching, West Germany	Peter Maier Komor H. J. Maier
October 1979	New England Nuclear Corporation, Boston, Massachusetts (USA)	Jozef Jaklovsky

At the Montreal, Canada, meeting, a proposal to form an International Nuclear Target Development Society (INTDS) was avidly accepted and officers and the Board of Directors were elected. Over the years the Board has had numerous changes in membership, but the President and Secretary were unchanged until the Boston meeting in 1979. From instigation these two officers were W. D. (Dan) Riel, President, and Joanne Heagney, Secretary. Mr. Riel is currently in charge of target preparation at the University of New York, Stonybrook, New York, and Ms. Heagney is co-owner (with her husband Joe) of a commercial target and special materials preparations firm known as Micromatter, Inc., Seattle, Washington (USA). In 1979 the author was elected President of the Society. It is to be noted that I.N.T.D.S. is incorporated as a non-profit organization under the state of California.

Generally, the International Nuclear Target Development Society appears to be a flourishing assembly which has answered the need for technology, publication and information transmittal, and international relations associated with the needs of target and special nuclear materials preparation. At present the Society membership spans 20 countries and includes 178 persons.

SOURCES OF SEPARATED ISOTOPES FOR NUCLEAR TARGETRY

E. Newman

Chemical Technology Division
Oak Ridge National Laboratory*
Oak Ridge, Tennessee 37830

INTRODUCTION

The use of enriched isotopes in all branches of nuclear research, including physics, chemistry, medicine, and engineering, has continued to develop and expand since they first became available. The specific benefit derived from the use of separated isotopes is, obviously, dependent on the nature of the experiment and on the particular effect under investigation. Enumeration of the varied applications of enriched stable and long-lived radioisotopes is beyond the scope of this article. However, in a broad sense, enriched isotopes increase the "signal-to-noise" ratio in many investigations and, in the case of isotopes with very small natural abundances, may be the only practical way of performing a meaningful experiment or obtaining a radioisotope with a high specific activity.

The increasing use of enriched isotopes in the preparation of targets for low-energy nuclear physics research may be illustrated by referring to publications in Nuclear Physics A. The data

*Operated by Union Carbide Corporation under contract W-7405-eng-26 with the U. S. Department of Energy.

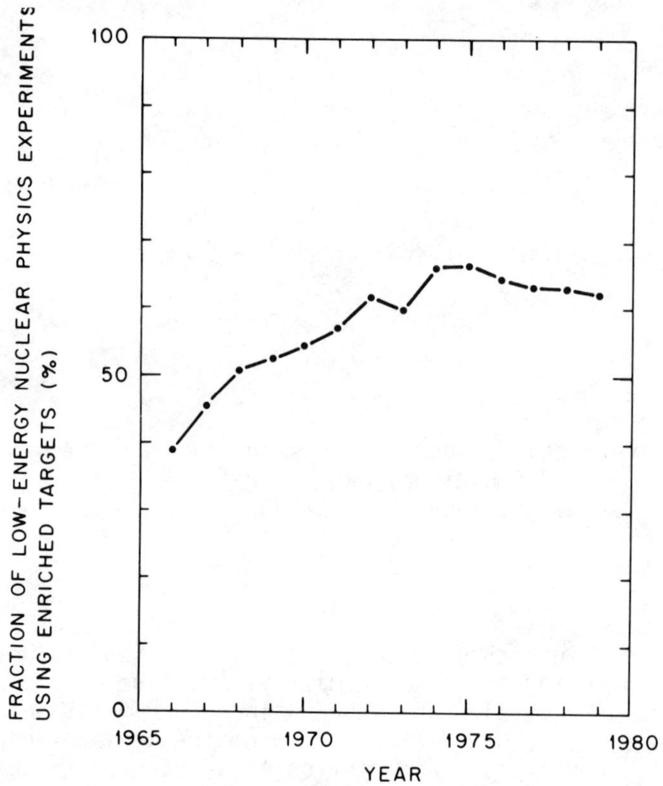

Figure 1. Percentage of low-energy nuclear physics experiments using enriched isotopes in the preparation of targets, as determined from publications in Nuclear Physics A during the period 1966-1979.

plotted in Figure 1 were derived from papers published in this scientific journal during the period 1966-1979. As shown by the figure, two-thirds of the investigations in this subfield now use enriched isotopes routinely.

The purposes of this review are to outline briefly the various methods of enriching isotopes and to discuss a few of the advantages and disadvantages associated with each process. Some methods will be omitted either because their applications are too limited or because they were found to be decidedly inferior to other enrichment processes.

ENRICHMENT METHODS

Electromagnetic isotope separation provides the single largest source of enriched isotopes for nuclear targetry. This method is characterized by large separation factors but relatively low throughput of material, while other separation methods provide large throughput but significantly smaller separation factors. The latter methods, notably diffusion, chemical change, and gas centrifugation, can be cascaded to provide high isotopic enrichment. The salient features of the principal isotopic enrichment processes are outlined in the following paragraphs.

Electromagnetic

The principal advantages of the electromagnetic approach are that the same apparatus and techniques are applicable to all elements in the periodic table, and that all isotopes of multi-isotopic elements are enriched simultaneously. The latter advantage is of considerable importance since in most of the other processes the enrichment is selective of either the heaviest or the lightest isotope of an element.

The production of separated stable isotopes in significant quantities requires a dedicated, high-throughput facility. The calutrons (i.e., very high-current mass spectrometers) at the Oak Ridge National Laboratory comprise such a facility. The devices, which were used to separate uranium isotopes in the 1940s, have been suitably adapted for the separation of isotopes throughout the periodic table.

The theoretical basis for the process is the motion of an energetic ion in a uniform magnetic field. As is well known, the solution of Maxwell's equations shows that the path of the ion is circular with the radius of curvature proportional to the square root of the mass divided by the electronic charge. Thus, there is a physical separation of all the isotopes of an element, and each can be collected individually and simultaneously.

The ORNL program will continue to be the major supplier of separated isotopes for research purposes in the foreseeable future.

Gas-Phase Thermal Diffusion

Gas-phase thermal diffusion is the main source of isotopes of the noble gases. The advantage of the process is that it can be cascaded and, with the appropriate number of stages, will produce high-assay product. The principal supplier is the Mound Facility in Miamisburg, Ohio.

The method is based on the fact that a temperature gradient produces a concentration or density gradient. Since it is known that density is approximately proportional to isotopic weight, the temperature gradient can result in an isotopic gradient and, hence, in isotopic separation. The temperature gradient in a thermal diffusion column is effected by having a center wire maintained at elevated temperature and a cooled wall. This gives rise to a radial distribution. Natural convection produces a countercurrent gas flow (i.e., down the wall and up the center). The difference in the isotopic concentration between the top and the bottom of the column determines the separation factor. The column approach lends itself quite well to a cascade to provide high-assay material. Although it can be successfully overcome, the principal disadvantage of this method of isotope separation is difficulty of enriching the "middle" isotopes.

Distillation

The low-temperature fractional distillation method of isotopic enrichment is utilized by the Los Alamos Scientific Laboratory (LASL) to provide stable isotopes of carbon, nitrogen, and oxygen. These products are regularly distributed by the Mound Laboratory in Miamisburg, Ohio.

Distillation separation takes advantage of the fact that isotopic species have different vapor pressures and boiling points. For example, the boiling point of D_2O is approximately 1.4°C greater than that of H_2O. There is also a small difference between the boiling points of $H_2{}^{18}O$ and $H_2{}^{16}O$ and between $D_2{}^{18}O$ and $D_2{}^{16}O$, but these do not interfere with the production of heavy water.

In the production of the carbon, nitrogen, and oxygen isotopes, LASL distills carbon monoxide and nitric oxide, at very low temperature, to obtain the enriched material. The former compound yields carbon, while the latter yields both nitrogen and oxygen. The high-boiling distillate is enriched in the heavy isotopes, and the individual elements are extracted from this fraction by standard chemical techniques.

Laser Isotope Separation

The use of lasers in the isotope enrichment field offers great potential since a very high separation factor can be achieved, and the technique can be used to either enrich or deplete one particular isotope of an element. The process also has the advantage that it is not power-intensive.

Two methods can be used to effect the separation of isotopes by lasers. The first involves separation at the molecular level.

The quantum levels of molecules are different, depending on the isotopic species present in the molecule. Thus, by using a monochromatic laser, certain of the particles in a gas or liquid can be selectively excited. The criterion for the selective excitation is that one absorption line of the desired species be shifted with respect to the absorption lines of the other species. Once the molecule has been excited, advantage is taken of its enhanced chemical reactivity to accomplish the physical separation.

The second method, which also depends on selective photoexcitation, is operable at the atomic level. The most promising method depends on photoexcitation followed by photoionization of a particular isotope and the deflection of the ion in an electrostatic field. This deflection allows a physical separation and, hence, the recovery of the enriched material.

Undoubtedly, the laser method of enrichment will continue to be developed for the foreseeable future. Detailed atomic and/or molecular spectroscopy data must be accumulated and evaluated before the process can be extended to most of the elements in the periodic table. The separation rate is a function of many variables, including the absorption cross-sections, available laser power, and allowable gas density, but large-scale operation is obviously feasible. Laser isotope separation will be applicable primarily in those cases where a particular isotope of an element is required.

Other Methods

Numerous other methods have been used for the enrichment of isotopes. Detailed discussions of these methods, which impact only marginally on the nuclear target community, are available in the literature; therefore, only a brief summary is given here.

Gaseous diffusion through a porous barrier, which is perhaps the best known method of isotopic separation, is primarily used to provide enriched uranium for nuclear fusion reactors. It is based on the theory that the rate of diffusion of a gas in a medium is inversely proportional to the square root of its density. Thus, since density is directly related to mass, a gas of lower molecular weight will pass through a porous barrier faster than one of higher molecular weight.

The use of centrifugal forces to separate isotopes has been reevaluated in terms of today's rotor technology and has been selected as the next-generation method of separating uranium. The forces generated in a rotating system produce a radial concentration with the heaviest mass at the largest radius. One notable advantage of the centrifugal method is that the separation factor is proportional to the mass difference rather than the ratio of the masses of the isotopes to be separated.

Two other isotopic enrichment methods deserve mention. The first is the electrolytic method primarily used for the production of heavy water. Electrolysis of aqueous solutions results in the preferential evolution of the light isotope, thus concentrating deuterium in the residual water. The second method, chemical exchange, takes advantage of the different chemical reactivities of the isotopes of an element. For example, the isotopic enrichment of the isotopes of nitrogen can be accomplished in a counterflow column of ammonia gas and ammonia nitrate. The heavy isotope tends to concentrate in the NH_4^+ ions in solution, while the light isotope concentrates in the gas phase.

SUMMARY

A variety of methods can be used to separate and enrich isotopes of interest to the nuclear targetry field. However, the production of significant quantities of these isotopes requires a dedicated facility with a large investment in capital equipment. Because of the unique requirements of the research community and their need for enriched samples of almost all the stable isotopes in the periodic table, the electromagnetic method of isotope separation will probably continue to be the primary source of material.

REDUCTION OF TiO_2, ZrO_2, AND HfO_2 FOR TARGET PREPARATION

Y. K. Peng

Tandem Accelerator Laboratory
McMaster University
Hamilton, Ontario, Canada, L8S 4K1

ABSTRACT

Using a small bomb type reaction vessel 50-100 milligrams of TiO_2, ZrO_2 or HfO_2 can be reduced to its metallic form at a temperature of $\sim$ 1000°C using magnesium as reducing agent. The metal obtained contains negligible amount of impurity and can be evaporated for preparing metallic targets. It, however, cannot be rolled to thin foils. The yield of this process ranges from 80 to 90%.

INTRODUCTION

Titanium, zirconium and hafnium metals react readily with oxygen at elevated temperatures and form very stable oxides. The reduction of these oxides cannot be achieved by usual methods such as the hydrogen reduction used in reducing iron oxide and lead oxide and the reduction-distillation used in reducing rare-earth metal oxides. A literature search has shown that these oxides, in a quantity of a few grams or more, can be reduced with calcium or magnesium metal at a temperature of $\sim$ 1000°C in an air tight reaction vessel under argon atmosphere[1,2,3].

This technique however cannot be applied directly to reduce the oxides of isotopically enriched materials. The amount of these oxides available is usually less than 100 milligrams, even a small amount of oxygen present in the reaction vessel could reduce the yield drastically. A minimum oxygen content and the cleaness of the reaction vessel and components used are therefore the prime requirements to be satisfied in reducing the oxides of isotopically

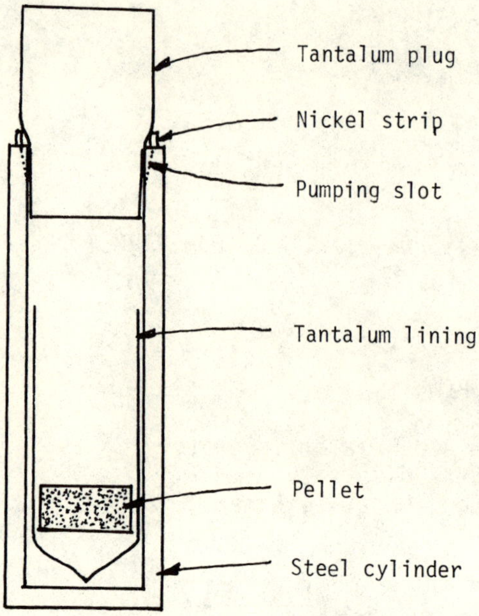

Figure 1. Reaction Vessel

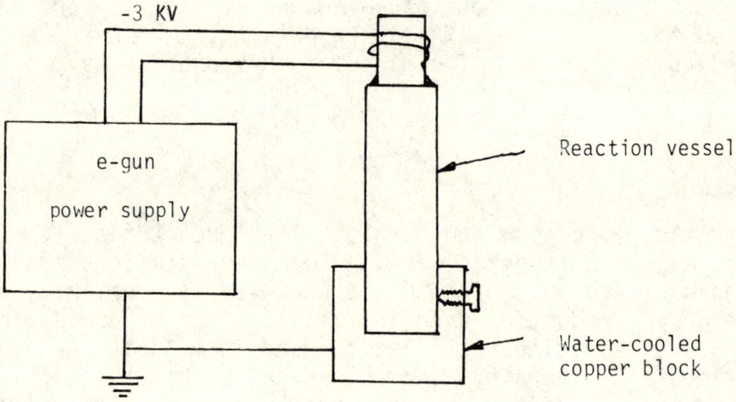

Figure 2. Mounting Arrangement

enriched materials. In this paper a modified technique which satisfies these conditions is described.

Method and Results

The equipment required for this process is a vacuum chamber, an electron-gun and power supply and a small tube furnace (or a small box furnace which could be modified to accept a small diameter quartz tubing). All this equipment is readily available in a target laboratory.

The assembled reaction vessel is shown in Figure 1. The mixture of 100 mg of oxide and ~60 mg of freshly filed magnesium powder of high purity is well mixed and compressed to a pellet. The pellet is placed in a tantalum lined steel cylinder and the cylinder is closed with a tantalum plug wrapped with a narrow nickel strip. At the joint two small slots are provided for pumping the air out of the cylinder.

The entire reaction vessel is placed in the evaporator and mounted on a water-cooled copper block as shown in Figure 2. The evaporation chamber is evacuated to a pressure less than 10^{-5} torr and after outgassing the reaction vessel briefly via electron bombardment, the joint is sealed by melting the nickel strip around the plug.

The completely sealed reaction vessel is removed from the chamber, placed in a continuously pumped tube furnace and heated to 1000-1100°C for 20 minutes. The charge usually remaining as a pellet is removed from the reaction vessel by cutting the vessel open, and treated first with diluted hydrochloric acid to remove magnesium oxide and any excess magnesium and then washed with deionized water. The black powder obtained, i.e., the metal, is dried, pressed to a pellet and melted down to a small shiny bead. The product yield of this process ranges from 80 to 90%.

The metal obtained can be evaporated for preparing metallic targets, but it cannot be rolled to thin foils. It, however, can be used as the starting material and be further purified to produce ductile metal through a van Arkel-de Boer purification process[4].

The oxides of isotopically enriched ^{47}Ti, ^{48}Ti, ^{49}Ti, ^{90}Zr, ^{178}Hf, and ^{180}Hf have been successfully reduced using this technique and targets of ~200 μg/cm² thickness have been prepared. Any impurity present in the targets was found to be negligible.

REFERENCES

(1) O. Ruff, Z. Anorg. Chem. 129 (1923) 267.
(2) J. H. de Boer and J. D. Fast, Z. Anorg. Chem. 187 (1930) 177, 198.
(3) W. Kroll, Z. Anorg. Chem. 234 (1937) 42.
(4) A. E. van Arkel and J. H. de Boer, Z. Anorg. Chem. 148 (1925) 345.

TARGET TECHNIQUES APPLIED TO γ-SPECTROSCOPY WITH HEAVY IONS

G. Sletten

The Niels Bohr Institute
University of Copenhagen
Denmark

The large energy losses and the straggling taking place when high energy heavy ion beams penetrate targets demand new target techniques. The large Doppler shifts resulting from emission of γ-rays in flight also become an important issue. Such shifts are wanted during lifetime measurements, but in most experiments the shifts are not wanted. The paper describes how these effects relate to the target maker and reviews experience with beams of O, Ar, Ti, Ni, Cu, Kr and Xe at energies of about 5 MeV/amu. The targets prepared for these experiments have been both backed and self-supporting with Z ranging from 6 to 70.

INTRODUCTION

Nuclear structure studies at high angular momenta involve the use of heavy ion accelerators and beams which deposit rather large energies in the targets. The large energy losses present difficult experimental situations, but also rather extreme conditions for the target foil itself. It is the purpose of the present paper to review some of the experimental problems and how these relate to the target design and fabrication.

Since the problems relevant to particle spectroscopy with heavy ions were reviewed by Erskine at an earlier INTDS meeting(1), only those problems which are special in γ-ray studies will be treated here.

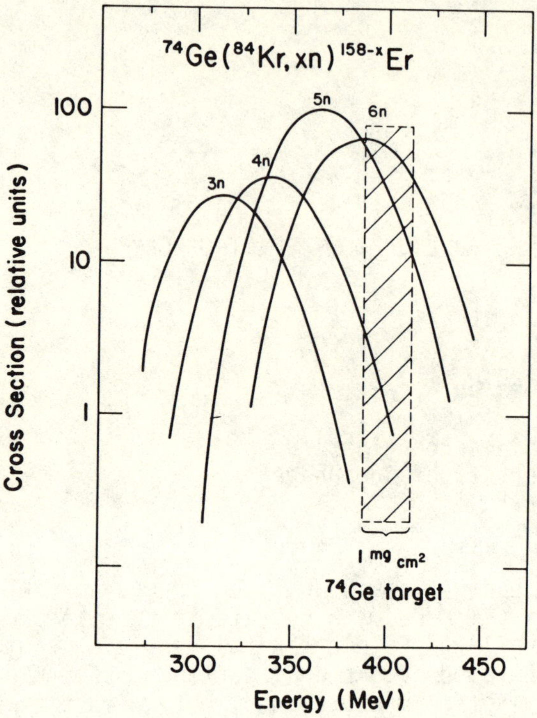

Figure 1. Excitation function for Er isotopes in Kr bombardment of a ^{74}Ge target.

Self-supporting Metallic Target Foils

A quite common experimental situation in heavy ion γ-ray spectroscopy is the use of about 1 mg/cm^2 self-supporting targets and beams of the order of 10^{11} particles per second.

There is nothing magic about the 1 mg/cm^2 thickness. In fact one would very often want it to be thinner, but for most of the rolling work we do in our laboratory, it is a practical limit.

The reason why one would like to have this thickness, or possibly thinner, is illustrated by Figure 1. The stopping of the beam in the 1 mg/cm^2 of target material reduces its energy by 25 MeV and introduces this as a $\pm$ 12.5 MeV uncertainty around a mean value of 400 MeV. The shaded area shows that the main products are evaporation of 5 and 6 neutrons. At lower incident energies the shaded box will be somewhat wider since the

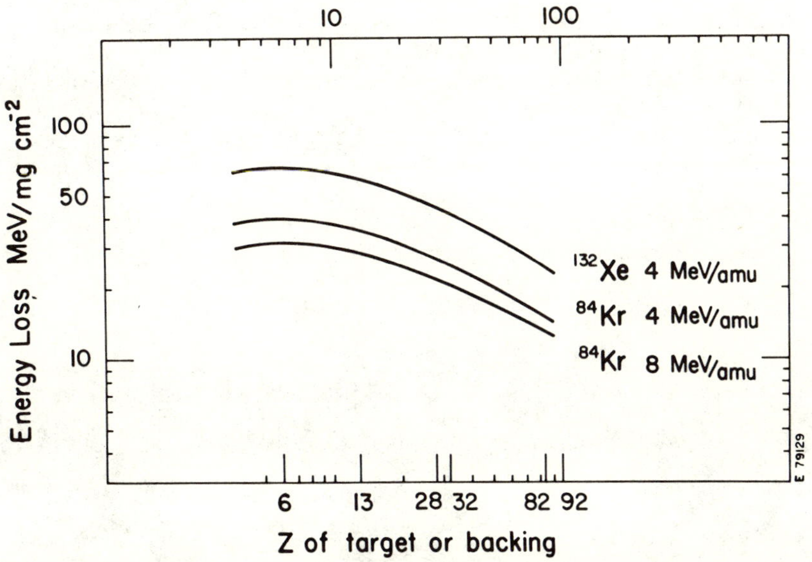

Figure 2. Energy loss of Xe and Kr ions in the elements. Stopping power data from Ref. 2.

stopping power increases with decreasing ion energy, Figure 2. The box will also cover 3-4 reaction channels, and the thickness of the target becomes of great importance with respect to channel selectivity.

There is nothing magic about the 10^{11} particles per second either, but it is chosen as an example. Whether it is realistic to use this intensity will be discussed below.

As an example for the discussion, let us consider bombardment of a 1 mg/cm^2 metallic foil of ^{96}Zr with 200 electrical nanoampères of ^{84}Kr ions at an energy of 4 MeV/amu. If the average charge of the ions is 12, this corresponds to about 10^{11} particles per second.

When the Krypton beam penetrates the 1 mg/cm^2 Zr target, the energy loss is about 25 MeV, Figure 2. This energy deposit can be expressed in joule/s or watts in the following way:

$$W = N \cdot \Delta E \cdot q,$$

where N is the number of particles per second, ΔE the energy loss in electron Volts and q the unit charge ($1.6 \cdot 10^{-19}$ Coulomb or J/eV). For the example chosen the energy deposited in the target is 0.4 watt.

The removal of this energy from a 2 mm x 2 mm beam spot takes place mainly by radiation since conduction of heat along a foil of actual thickness $\sim 10^{-4}$ cm (1 mg/cm^2) is extremely small. Under this assumption one can compute the temperature of the beam spot, treating both the target and the surroundings as black bodies[3].

With a beam spot size of 4 mm^2 and therefore a deposit of 10 watt/cm , an equilibrium temperature of about 900°C will be

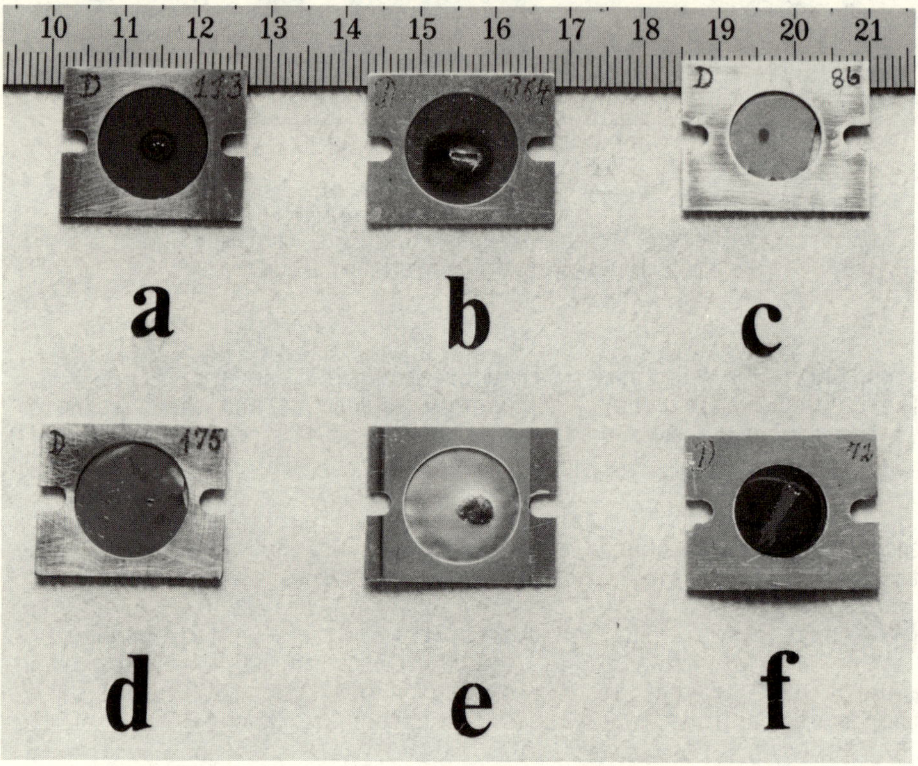

Figure 3. Targets after exposure to heavy ion beams.

a) 3.8 mg/cm^2 Zr foil with 200 nA ^{84}Kr.
b) 1 mg/cm^2 ^{64}Ni foil with 200 nA ^{84}Kr.
c) 1 mg/cm^2 ^{122}Sn foil with 15 nA ^{50}Ti.
d) 1 mg/cm^2 ^{124}Sn foil with $\sim$ 15 nA ^{50}Ti.
e) 1 mg/cm^2 ^{124}Sn on 12 mg/cm^2 ^{208}Pb with 10 nA ^{40}Ar.
f) 1 mg/cm^2 ^{76}Ge on 12 mg/cm^2 Au with about 30 Na ^{86}Kr.

For details, see text.

reached under these experimental conditions. This is well below the melting point of Zr at 1852°C, and evaporation only becomes important beyond 1800°C.

It is clear from the example chosen that there are no experimental problems as far as the stability of this target is concerned. One could ask within which limits is the target stable? From Reference 3 one finds that the beam spot reaches the melting point of Zr at an energy deposit of about 110 watts/cm^2. At the same temperature the vapor pressure of Zr approaches 10^{-4} torr, and the target would certainly perish. This would, however, require a beam intensity 10 times larger which is probably not obtainable.

With a smaller beam spot area, however, a smaller beam current is required to reach the critical equilibrium temperature. The extreme of 500 nA through a spot of 1 mm^2 would drive the temperature to 1800°C and probably ruin this target.

Figure 3a actually shows an example of a metallic Zr target with a melted hole in it. Here the target thickness was 3.8 mg/cm^2, and the energy deposit correspondingly larger, 1.52 watt, or 38 watt/cm^2 over the about 4 mm^2. This would normally keep the target 500°C under the melting point. The apparent $\sim$ 1 mm^2 hole probably results from a temporary change of focus giving rise to an energy deposit of 152 watt/cm^2 and a corresponding equilibrium temperature of about 2000°C, well above the melting point of Zr.

From the discussion above it is clear that tables of stopping power are indispensable for target makers supplying targets to heavy ion experiments. Some of the general features of such tables are illustrated in Figure 2. For a given ion and energy the stopping power has a 1/Z dependence, and for a given beam and Z there is a 1/E dependence. It is also seen that the energy loss increases with the Z of the beam particle, and if we in the example of the 1 mg/cm^2 Zr target let the beam be 10^{11} particles $\cdot$ S^{-1} of ^{132}Xe, the energy deposit would be 16 watt/cm^2 as compared to 10 watt/cm^2 for ^{84}Kr. It should be noted that the electrical current of Xe will be higher because of the higher charge per particle.

A ^{64}Ni target 1 mg/cm^2 thick shown in Figure 3b has also been destroyed in beam. Here the target was designed for a beam spot size of 0.04 cm^2 and 200 nA of ^{84}Kr beam which would keep the heat deposit down to 11·2 watt/cm^2 corresponding to an equilibrium temperature of 920°C well below the melting point at 1453°C. Inspection of the target clearly indicates melting which could have happened during refocusing of the beam through sizes of about 1 mm^2 and driven the temperature beyond 1400°C. Both targets mentioned here are of elements which are considered as very good target materials both with respect to mechanical and thermal properties.

Nevertheless, serious damage has been observed. How then can targets that are not of such noble metals survive in beam?

One obvious thing to do to reduce target destruction would be to use less beam. In several experiments, however, the cross-sections are low and the highest possible beam current is demanded. Therefore a large area beam spot with homogeneous intensity distribution would help. Other remedies might be a target wobbler or even a beam wobbler.

With a melting point of only 232°C it seems, after the preceding discussion, almost impossible to use a tin target in a heavy ion beam. Figure 3c shows a ^{122}Sn target, $\sim$ 1 mg/cm^2, which has been exposed to a ^{50}Ti beam at 240 MeV. The energy loss of the Ti beam is 11 MeV/mg cm^{-2}, and we have found by bitter experience about 15 electrical nA to be the destruction limit with a beam spot of about 4 mm^2. With the same assumptions as in the case of the Zr target and with a charge of about 10 for the Ti ions, an equilibrium temperature of about 250°C is obtained. This is above the melting point of tin and the reason why the target survived is not clear. At any rate the 15 nA is at the very limit of what the target can tolerate. A close look at the surface of Figure 3c reveals a very shiny replica of the beam of somewhat different structure than the rest of the rolled foil. This indeed suggests that the foil has been at the melting point.

Figure 3d shows a ^{124}Sn target of the same thickness where a refocusing of the beam resulted in a smaller beam spot, higher energy deposit per unit area and melting.

Targets on Backing Foils

It has been possible to prepare the targets discussed so far as self-supporting foils. The use of backing materials for those elements that cannot be prepared as self-supporting foils introduces problems of the same kind as discussed above.

With respect to energy loss in the backing and thereby creation of heat, Figure shows that one should select as heavy Z material as possible in order to minimize the temperature for a given backing thickness. The use of high Z backing has the further advantage that the Coulomb barrier for nuclear reactions diminishes with increasing Z. Therefore a high Z backing reduces the unwanted background to a minimum.

Our experience with heavy ion induced reactions at about 5 MeV/amu and lower is that Au and Pb are good backings. Both can be produced in a range of thickness from about 0.1 mg/cm^2 and upwards with excellent mechanical properties. Heavy ions of

5 MeV/amu are below the Coulomb barrier for these elements, and the only background created is the γ-radiation from Coulomb excitation.

This background is, however, substantial for Au, but quite small for Pb. Especially, ^{208}Pb is preferred if one wishes to reduce the background to a minimum. To take full advantage of the ^{208}Pb, the backing should be free of oxide layers since oxygen may react with the beam particle and create background radiation.

Our favorite backing is therefore ^{208}Pb, but alas, also this has its deficiencies. Lead melts at 328°C and in view of the discussion above this puts limits on the amounts of beam that can be tolerated.

The Effects of Doppler Shifts on Target Design

Doppler shifts are phenomena that require much attention in γ-ray spectroscopy with heavy ions and impose a variety of problems on the target makers.

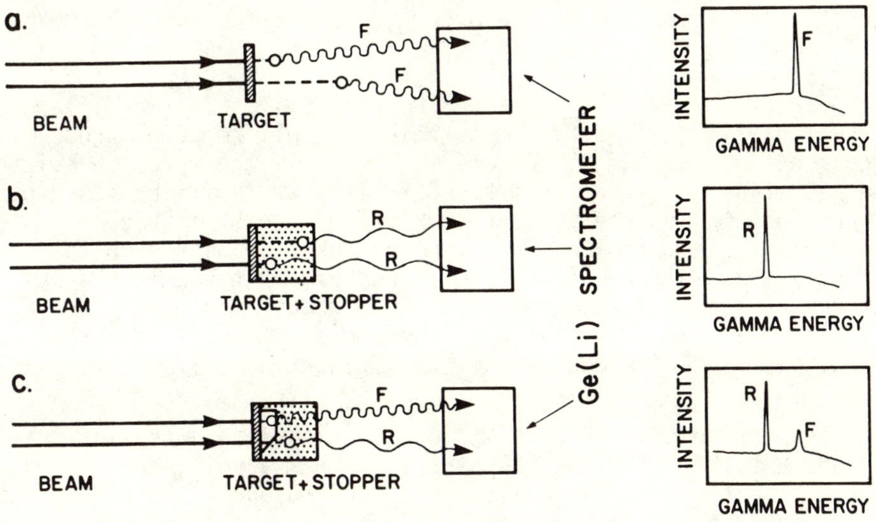

Figure 4. The Effect of Doppler Shifts on γ-ray spectra. F designs γ-rays emitted in flight and R γ-rays emitted at rest.

a) Unbacked, thin target. b) Thin target with stopper. c) Thin target with voids between target and stopper.

Doppler shifts arise because the source emitting the γ-rays moves with respect to the detector. The shift, ΔE_γ, and the transition energy E_γ are related by:

$$\Delta E_\gamma = E_\gamma \beta_r \cos\theta, \qquad (1)$$

where θ is the angle between the source trajectory and the direction of the γ-ray propagation, and β_r is the ratio of the source velocity to the velocity of light.

Ratios of a percent and more are customary in nuclear structure studies, and one is often dealing with shifts of the order 10 keV for a 500 keV γ-ray at θ = 0°.

The lifetime of the state emitting the γ-ray and the stopping time for the recoiling nucleus are two decisive factors when discussing Doppler shifts. It is clear that if the stopping time is short as compared to the lifetime of the emitter, β_r vanishes and the shift, ΔE_γ, is zero. Therefore, a medium offering the shortest possible stopping time is wanted, when no Doppler shift is wanted.

Both Au and Pb are good stopping materials in this connection, and stopping times are of the order 10^{-13} sec for about 70 MeV rare earth recoils.

Figure 4 shows the effect of Doppler shifts in two different situations. In a) the target is thin, and all reaction products recoil out of the target and are moving while the γ-ray is emitted. The resulting γ-ray spectrum at 0° from decay in flight has a peak F corresponding to the nuclear transition. In b) a thick layer of a stopping material, i.e. Au or Pb, has been deposited on the target so that all recoils are stopped before they emit a γ-ray. The resulting γ-ray spectrum at 0° from decay at rest has a peak R which then is the true energy of the nuclear transition. The energy peaks R and F equals the ΔE_γ of equation (1).

In section c) of Figure 4 the situation in Figure 4b is repeated, but this time the stopping layer has a void, and the resulting spectrum has two peaks since a fraction of the recoils now decays in flight. In order to have well defined conditions it is therefore imperative to have good adherence between the stopping layer and the target itself. Evaporation or sputtering techniques are recommended in these preparations. Sandwiching by rolling is too hazardous because of the possibility of pocket formation.

The necessary thickness of the stopping layer depends on the recoil velocity. Rare earth recoils of about 75 MeV require 10-15 mg/cm^2 of Au as the stopper. Thus the thickness of the stopper exceeds that of the target itself by an order of magnitude. It is therefore the energy loss of the beam particles in the stopper that determines what beam intensity one can use.

Figure 3e shows a target of 1 mg/cm^2 ^{124}Sn evaporated onto a 12 mg/cm^2 ^{208}Pb stopping foil. It has been subjected to ^{40}Ar beams of about 180 MeV, and the destruction is clearly visible. The beam intensity has been about 10 electrical nA over an area of 4 mm^2 and the deposited energy therefore 2 watt/cm^2. Taking only radiation into account, a temperature of 500°C is reached which exceeds the melting point of both Sn and Pb and induces a vapor pressure of Pb up to 10^{-4} torr.

Figure 3f shows a 1 mg/cm^2 ^{76}Ge target deposited on a 12 mg/cm^2 Au stopper which has been subjected to a ^{86}Kr beam. Here the beam intensity has been between 20 and 30 electrical nA over an area of 4 mm^2. An intensity of 25 nA would bring the temperature of the beam spot close to the melting point of Ge, 937°C.

The vapor pressure of Ge at 950°C is 10^{-6} torr and 10^{-4} torr at 1170°C. It is believed that the destruction of this target took place by melting and development of a pocket of Ge vapor at the Ge-Au junction which then burst.

The possibility also exists that the lifetimes are extremely short so that whatever the stopper is made of, the nucleus moves during emission. In this case one would therefore prefer to have all recoils moving in order to observe the full Doppler shift at 0°. The target would in this case be the thinnest possible self-supporting target.

One can see from eq. (1) that β_r and therefore the recoil velocity is proportional to ΔE_γ. In order to reduce the spread in ΔE_γ it is therefore important to have a sharpest possible velocity distribution. To obtain this one must demand a high degree of target uniformity and use the thinnest targets possible.

So far the discussion of Doppler shifts has only treated unwanted shifts. For purposes of lifetime measurements the shift is often referred to as the plunger method. In such experiments one would like thin targets with perfectly plane surfaces in order to obtain a small velocity spread of recoils and a sharp time-equals-zero definition. Otherwise, the general discussion of this paper is relevant also to this experimental technique.

Conclusion

Our experience with heavy ion bombarded foils at about 5 MeV/amu is that the large energy losses present extreme conditions to targets of all elements. Even target foils of zirconium might perish under bombardment of a 200 nA krypton beam.

It is found that the heat deposited in the target spot must be radiated away and that heat conduction through the foil is negligible. The beam spot therefore reaches its equilibrium temperature almost immediately, and the only way to avoid destruction of the target is adjustment of beam spot size and beam intensity. For targets of elements with poor thermal properties a continuous movement of the target or wobbling of the beam might give additional safety.

The large Doppler shifts observed in work with heavy ions can in most cases be removed by the use of appropriate stopping media. The large thicknesses needed in order to stop the recoils emitting γ-rays impose new heat problems at the target spot. Extreme care with respect to beam intensity must be taken in such experiments to avoid pocket formation and melting of the stopper.

ACKNOWLEDGEMENTS

It is a pleasure to acknowledge the devotion and skillful assistance of Mrs. Sonja Dahl and Mrs. Jette Sørensen in preparing the targets described here. Especially discussions of the destructive effects of beam are appreciated.

REFERENCES

(1) J. R. Erskine, Anl./Phy./MSD-76-1, p. 141. Fourth Annual International Conference of the Nuclear Target Development Society, Argonne, Ill., U. S. A. (1975)
(2) L. C. Northcliffe and R. F. Schilling, Nuclear Data Tables A7, 233 (1970).
(3) Handbook of Tables for Applied Engineering Science. The Chemical Rubber Co., Cleveland, Oh (1970), p. 448, 449.

EXPERIENCE WITH THIN HAVAR FOILS FOR CYCLOTRON TARGET WINDOWS

Lester S. Skaggs

King Faisal Specialist Hospital and Research Centre
Riyadh, Saudi Arabia

and

Franca T. Kuchnir, Frank M. Waterman*, and Helmut Forsthoff

Franklin McLean Memorial Research Institute**, and Department of Radiology***, University of Chicago, Chicago, Ill.

Thin foils of Havar† have frequently been used as windows to separate a high pressure gas target from the vacuum system of the accelerator. This cobalt-based alloy is non-magnetic, corrosion resistant and has unusually high mechanical strength and resistance to fatigue. Tensile strength as high as 2.5×10^9 for the cold reduced and heat treated alloy have been reported.

---

* Current Address: Department of Radiation Therapy and Nuclear Medicine, Thomas Jefferson University, Philadelphia, PA 19107.

** Operated by the University of Chicago for the U. S. E. R. D. A. under Contract No. EY76-C-02-0069.

***This work supported in part by the Frederick E. and Ida H. Hummell Foundation and by Grant #78-60 from the American Cancer Society, Illinois Division, Inc.

† Havar, an alloy supplied by Hamilton Technology, Inc., P. O. Box 1609, Lancaster, Penn. 17604.

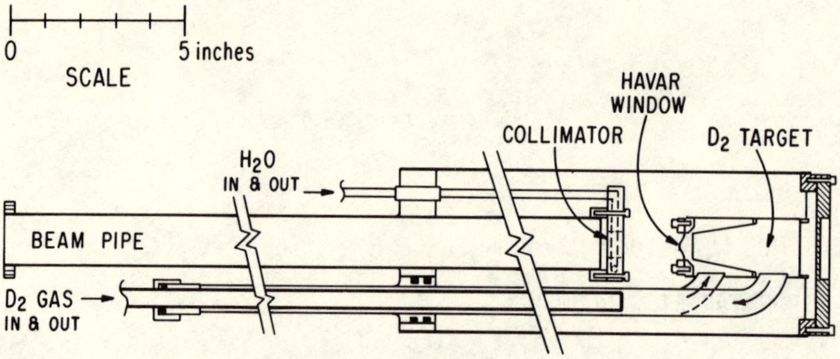

Figure 1. Cross-section of the cryogenic deuterium gas target vessel and enclosing dewar.

INTRODUCTION

The experience reported here was obtained with a deuterium gas target[1,2] used at the University of Chicago to produce a neutron beam for radiation therapy. The extracted deuteron beam of 8.3 MeV energy from the cyclotron[3] was transported some 8 meters to the target in a beam line with two quadrupole magnets. The target was operated at liquid nitrogen temperature, 77°K, and 10 atm. pressure. Figure 1 shows the target with the adjacent portion of the beam line and an evacuated Dewar for thermal insulation of the cold target vessel. The beam enters the system from the left, passes through the Havar window and is totally stopped in a target vessel of only 8.3 cm depth. Cold circulating gas enters the vessel through one of two transfer lines, is caused to blow across the window by the conical partition and then exits at the back. The gas is cooled in a heat exchanger immersed in liquid nitrogen in a Dewar which also contains a recirculating pump. The system is designed to dissipate 1500 watts of beam power (180 μA at 8.3 MeV) with only a 10°K rise in gas temperature in the target vessel.

A window of 1.9 cm diameter was chosen as a reasonable compromise between a small diameter for considerations of strength and a large diameter for ease of beam transport. Previous experience had indicated a foil of 1.27×10^{-3} cm thickness would have adequate strength. However, a test jig with a 1.9 cm dia. window was prepared and bursting test run with commercially annealed foil. At room temperature, the foil burst at 15.0 atm., at 77°K it burst at a surprising 21.8 atm. Thus the foil has a 50% safety factor at room temperature and over 100% at 77°K.

Aside from minor problems with the gas recirculating pump, the major problem with target operation has been with the Havar window. In the initial use it was found not possible, with any adjustments of the quadrupole currents, to put more than 40 μA of the 180 μA design current through the foil without rupture. It is well known that a cyclotron beam is likely to have hot spots in its distribution and these might be a possible cause of early failure. A possible solution considered was spreading the beam with a scattering foil. Such spreading produces a gaussian distribution of the beam and it has been shown[4] that, unless a large fraction of the beam is thrown away, this can produce a temperature almost double that for a uniformly distributed beam. Another option is rotation of the beam about the center of the window. This may produce a "double gaussian" (a gaussian distribution offset from and rotated about the center of the foil) but the above reference has shown this to be better than the single gaussian of the scattering foil. In addition, rotation of the beam by deflection in a rotating magnetic field allows more convenient control of the beam parameters.

Rotation of the Cyclotron Beam

Benaroya and Ramler[5] have described a motor-stator like magnet for deflecting and steering an external accelerator beam. Some installations have actually used A.C. motor stators for this purpose although the latter do not usually have the most favorable winding distribution. The core of the Benaroya and Ramler type magnet, being of solid steel, would not be suitable for rapid rotation. But a commercial 3 phase motor stator, constructed of laminated electrical steel, can be procurred, and rewound to produce a rotating magnetic field with 3 phase excitation. The stator obtained for this purpose had an inside diameter of 8.9 cm - sufficient to easily accommodate the beam line - and a length of about 10 cm. When placed at a distance of 2 meters from the target, such a magnet would deflect the beam 5 mm with a field of only 150 gauss. The stator had 36 slots and the necessary turns in each slot were calculated using the well solenoid equation, $H = 4\pi nI$, where H is the field in gauss, I the current in amperes and n the number of turns per cm of gap height in the field.

Each of the 3 coils were wound in two equal sections distributed about a symmetry plane containing the axis of the stator and passing midway through the steel lips between two adjacent winding slots. The gap height was calculated from the stator inner radius and the sine of the winding slot interval, 10°. From the solenoid equation above this gave 123 ampere turns for the coil in the first pair of slots on one side of the symmetry plane. For each succeeding pair of slots, the gap height was calculated from the difference of the sine of each additional

10° angle less that of the previous angle. The number of ampere turns thus decrease in each pair of slots going away from the symmetry plane, but in such a manner that the planes of constant magnetomotive force are essentially parallel to the symmetry plane and the gradient between planes is constant. Under these conditions the magnetic field for a given coil (consisting of identical halves on each side of the symmetry plane) is constant throughout the stator opening except for fringing fields at the end. These latter have only a second order effect in a particle deflecting magnet. The coils for the other 2 of the 3 phases were wound in an identical manner but were each rotated 120° about the axis from the first coil. Choosing the order of 8 amperes as a reasonable current, 123 ampere turns required 15 turns in the pair of slots nearest the symmetry plane. Each of the completed coils was tested with 8 amperes D. C. and a G. E. flux meter showed a field of 200 gauss. Searching the field showed it to be quite uniform throughout most of the space in the stator.

The magnet was easily operated from the 208/120V three phase, 60 cycle lines using a 3-phase Variac to adjust the current. With the magnet adjusted to produce a deflection of 0.5 cm, the maximum beam current was increased to 70 μa. This was almost double that obtained with a stationary beam, but less than half the design value. The instantaneous heating of the foil was then investigated. Assuming the current could be concentrated by the quadrupole into a 2 mm by 2 mm spot, the temperature in the foil would rise at the rate of 2 x 10^5°C/sec. With 60 cycle rotation in a 1 cm dia. circle, the beam would produce a temperature excursion of 200°C.

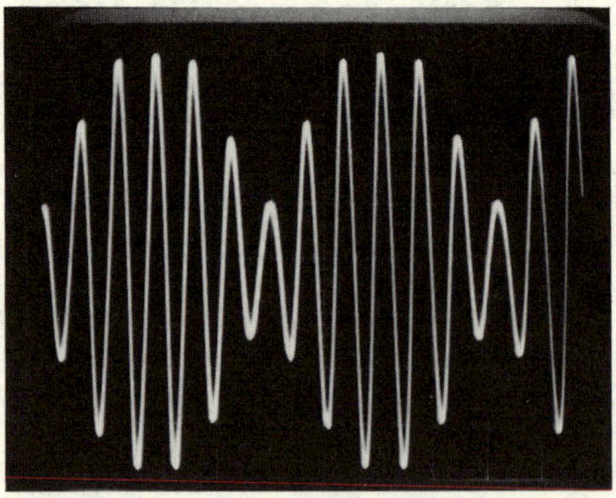

Figure 2. Oscilloscope trace of the driving voltage on one phase of the magnetic beam rotator at 715 Hz 50V/Div. vertically; 2 x 10^{-3} sec/Div. horizontally.

This could be too large an excursion on top of an already hot foil. It could be reduced to a moderate value by increasing the magnet excitation and rotation frequency by a factor of 10.

A three phase, variable frequency A. C. power supply was obtained consisting of an oscillator driving three audio-type solid state power amplifiers. It was modified by connecting an analog multiplier between each of the three outputs of the oscillator and each amplifier input. The second input of the multipliers were all three driven in common from a full wave rectified but unfiltered 60 cycle A. C. voltage with a D. C. bias in the same direction as the A. C. rectified voltage. This resulted in a modulation of the current to the magnet causing the beam on the foil to spiral from a minimum diameter determined by the D. C. bias to a maximum diameter determined by the sum of the D. C. bias and the rectified A. C. voltage. An oscilloscope trace of the voltage supplied to one phase of the magnet is shown in Figure 2. The result of this form of magnet excitation was that the beam was spread over a wider band on the foil even if it were sharply focused. The wave from the rectified A. C. was such that the beam spent less time at the inner diameter where the beam current density would otherwise be higher and more time at the maximum diameter. No attempt was made to synchronize the higher frequency with the 120 cycle modulation frequency since a non-integral ratio of the two would result in more even heating of the foil. Since it was known that the cooling of the foil by the cold deuterium gas would be a linear function of the area heated on the foil, it was expected that both the modulation and the higher rotational speed would contribute to larger beam currents through the foil.

In order to present a more favorable load to the solid state amplifiers with a power factor near unity, the magnet was operated with capacitors in parallel with each coil to tune the coil to a frequency of about 800 Hz. A small amount of resistance was placed in series with each coil to make the tuning less critical. At 800 Hz and 150 gauss the voltage across each coil was about 130 volts, rms, the circulating current about 6 amperes and the current from the power supply about 2 amperes in each phase.

In retrospect it has been realized that a more economical magnet for high frequency operation could be built using a 2 phase design with the phases at 90° on the stator. The high frequency power supplies are available for 2 phase, 90° operations and require only two power amplifiers and two multipliers for modulation. The A. C. current in each coil needs to be slightly larger, but this is not a critical factor.

No problem was experienced with heating of the stator or the windings even though it was run up to 800 Hz and the motor was

Table I. Life Experience of Havar Windows

Foil #	Use Period (1979) from:	to:	"Beam On" (120μA) (hours)	Remarks
9	2/20	3/6	10.4	Preventive Change
10	3/22	4/3	12.4	Blow Up
11	4/7	5/22	17.7	Slow Leak
12	5/22	5/24	3.7	Slow Leak
13	5/29	6/16	9.3	Slow Leak
14	6/19	7/2	2.0	Preventive Change
15	7/5	7/19	2.7	Blow Up
16	7/24	9/24	7.5	Blow Up

only designed for 60 Hz. This could be expected since the maximum field in the air gap was 150 gauss and concentration in the iron of the stator would not increase this by more than a factor of about 5. Heating of the beam tube was at the tolerable limit in a 5 cm diameter by approximately 0.25 cm wall of stainless steel. This problem could be easily solved with a ceramic section of beam tube coated with aquadag to dissipate electrical charges or possibly even with the stainless steel tube with thinner walls.

The results with the modulated fast-rotating magnet have been very gratifying. Beam currents of 120 μA have been consistently run with good foil lifetime. Table I shows the foil experience over the first 5 months after the new system was installed. The cause or the location of the slow leak which resulted in some foils being changed has not been determined though it does not seem to be a case of gas diffusing through a hot foil. The foils are sealed with indium wire o-rings and these might well be suspect. Since the beam is totally stopped in the deuterium gas and it is not heated to more than 88°K, it is not possible that the target vessel and indium seal get too warm. The energy lost in the 1.27×10^{-3} cm thick foil by the 8.3 MeV deuteron beam is about 0.6 MeV and 120 μA of beam current means a foil dissipation of 72 watts. It was not possible to observe the foil directly under beam bombardment to see if it was hot enough to glow. Havar foils 2.54×10^{-3} cm thick have been reported to hold 6 atm. pressure over a 3.8 cm dia. window while glowing brightly enough to be seen in a closed circuit T.V. system(6). It was possible

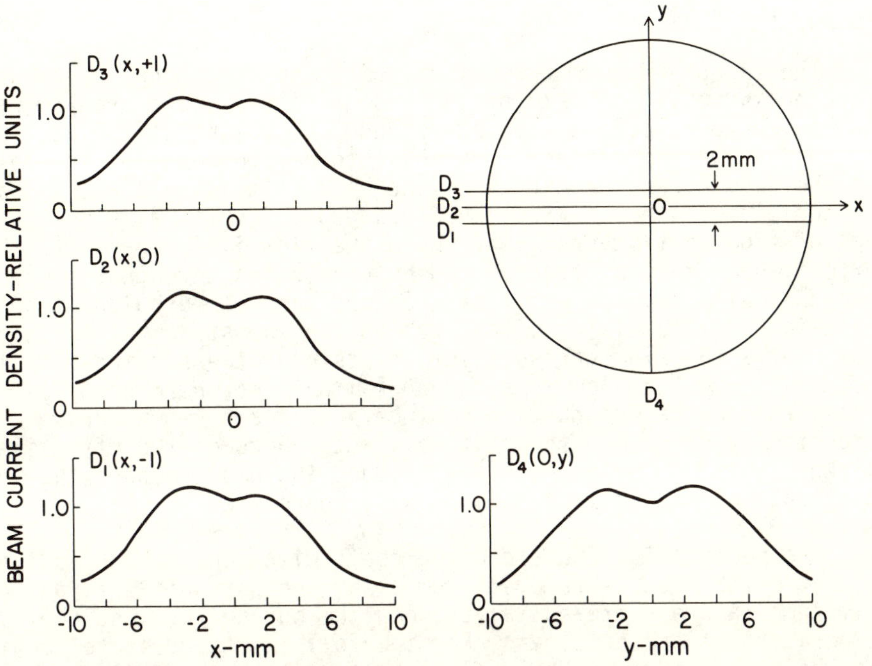

Figure 3. Plot of cyclotron beam current distribution on the window measured with a radioautograph of a Havar foil after several hours use. Three horizontal scans on left; vertical scan on right.

to observe with a T. V. system a grid of 1.27×10^{-3} cm dia. glowing tungsten wires placed a short distance in front of the window and these showed a nebular annular ring of beam. The center of the ring was rather well filled with beam and the wires in this region and immediately outside the ring were only slightly less bright than the area in the ring. This was expected since the cyclotron beam cannot be focused to a point or even to a sharply delineated area. A radioautograph was made of one of the foils after it had been used for some time. Figure 3 shows several densitometer scans across the radioautograph film. Since the film exposure was primarily to beta rays and without intensifying screens, the traces should be a reasonably linear representation of beam current. The beam distribution is "double gaussian" and not much is evidently being lost on the collimator or window frame at a radius of 9.5 mm.

Cooling the Foil

The 120 μa of beam current at which the target has been operated is still only 2/3 of the design value. So a program was undertaken to calculate and understand the cooling of the foil by the cold deuterium gas and, possibly, to devise means of increasing the beam current capacity of the foil. It is well known that the cooling of a hot surface by a gas depends critically on the type of flow that occurs[7,8]. When the flow is laminar past the surface, heat must be transmitted by conduction through the gas into the mainstream of gas flow. In the case of turbulent flow there is complete mixing of hot gas in the mainstream of turbulent flow, but a thin laminar sublayer exists near the stationary solid boundary where the flow is slow and laminar and where heat must be transmitted by gas conduction. Since values are known for conductivity of gases, the critical factor becomes the effective thickness of the laminar sublayer. Once the heat reaches the turbulent region, it is rapidly mixed and removed.

One needs to define a cooling coefficient, h_c, which is the rate at which heat is removed from a surface per unit area and per unit temperature differential between the hot surface and the bulk temperature of the cold gas. For the foil in steady state condition heat removed must be equal to heat lost in the foil by the beam and the cooling coefficient becomes

$$h_c = \frac{i \times \Delta E}{A(T_f - T_g)} \quad \frac{\text{watts}}{\text{cm}^2\ {}^\circ\text{K}} \qquad (1)$$

where

i = Beam current, μA
ΔE = particle energy lost in foil, MeV
A = cooled area of foil, cm^2
T_f = foil temperature, °K
T_g = bulk temperature of gas, °K

The basic problem is determining a value for h_c. This will depend on the character of gas flow past the foil and, if it is turbulent, on the degree of turbulence. It will be assumed that the flow is turbulent and characteristic of a long circular duct and shown later why this must be true.

Numerous experiments have shown that for a gas the effective thickness of the laminar sublayer is primarily a function of a dimensionless quantity known as the Reynold's number. For forced convection in long circular ducts of diameter, D, it has been shown that the following relation holds

$$(Nu) = 0.023\ (Re)^{0.8}\ (Pr)^{0.33} \tag{2}$$

where

$$(Nu) = \frac{h_c\ D}{k} \tag{3}$$

and

(Nu) = Nusselt number, dimensionless
(Re) = Reynold's number, dimensionless
(Pr) = Prandtl number, dimensionless
k = gas conductivity, watts/cm°K
h_c = cooling coefficient, watts/cm^2 °K

Since equation (2) is dimensionless in all terms, any exponents may be chosen that fit the experimental data. The value of the equation lies in the fact that the form and the exponents given above fit the experimental results over a wide range of turbulent flow in ducts of various sizes and fluids of different states and many different kinds. Equation (3) defines the Nusselt number. Substituting equation (1) in equation (3) and rearranging, one obtains

$$\frac{i \times \Delta E}{A} = k\ \frac{T_f - T_g}{D} \times (Nu) = k\ \frac{T_f - T_g}{D/(Nu)} \tag{4}$$

The more common interpretation of the Nusselt number is given by the first form of the last equation where it is seen that it is the ratio of the actual temperature gradient at the wall (foil) to the gradient that would exist over a distance D in the gas. In the second form of the equation, it is seen that D/(Nu) gives the effective thickness of the sublaminar boundary layer. This divided into the temperature difference between foil and gas gives the temperature gradient by which the gas conductivity factor must be multiplied to obtain the power removed from the foil per unit area. Since the temperature of the gas is known (it rises only ten degrees Kelvin when the total beam is stopped), the temperature of the foil can be determined. It remains only to evaluate the other two dimensionless numbers and to show that the equations for the long duct apply. First the numbers will be discussed and evaluated for the observed conditions which were a flow of 1.8 liters/sec of deuterium gas at 78°K and 10 atmospheres pressure. The Reynold's number is

$$(Re) = \frac{V \times D \times \rho}{\mu} \tag{5}$$

where

$$V = \frac{1800 cm^3/sec}{\pi(0.9525)^2 cm^2} = 632 \frac{cm}{sec} \quad \text{- gas flow velocity}$$

D = 1.905 cm - duct diameter
ρ = $6.29\ 10^{-3}$ gm/cm^3 - density of D_2 at 78°K & 10 atm
μ = $5.01\ 10^{-5}$ $\frac{gm}{cm\ sec}$ - viscosity of D_2 at 78°K & 10 atm

then

$$(Re) = 1.50 \times 10^5$$

Below a Reynold's number of 2000, the flow is usually laminar; between 2000 and 13000 it is usually turbulent; above about 13000 it is always turbulent and for the cryogenic target here considered it was far in the turbulent region. The physical significance of the Reynold's number can be seen from the following rearrangement of equation (5) and multiplication by V/V = 1

$$(Re) = \frac{\rho V^2}{\mu V/D}$$

The numerator is the flow of momentum through a stream tube of unit cross-sectional area and may be taken as representative of the inertial force per unit area. The quantity V/D is proportional to the velocity gradient across the tube or duct and when multiplied by the viscosity it is representative of the viscous shear stress. The Reynold's number is thus a measure of the ratio of inertial forces to viscous forces. Therefore when the Reynold's number is less than 2000, the dissipative viscous forces are predominant and any incipient turbulence is rapidly damped out. On the other hand, when the Reynold's number is above 2000 and certainly at 13000 or above, the inertial forces are predominant and the turbulence is certain to build up.

The other dimensionless factor in the expression evaluating the Nusselt number was the Prandtl number. It is given by

$$(Pr) = \frac{C_p \mu}{k} = \frac{\mu/\rho}{k/\rho C_p} \tag{6}$$

where μ, ρ and k have been defined and

C_p = 1.763 $\frac{cal}{gm°K}$ - specific heat at constant pressure of D_2 at 78°K & 10 atm

For deuterium at 78°K and 10 atm, $k = 41.6 \times 10^{-5}$ W/cm°K, values of ρ and μ have already been given and it is found that

$$(Pr) = 0.89$$

The Prandtl number is a characteristic of the fluid and is only slightly smaller than one for all gases. From the second form of the equation it may be seen to be the ratio of the molecular diffusivity of momentum to the molecular diffusivity of heat.

The expression for the Nusselt number in terms of the Reynold's and Prandtl numbers can now be evaluated from equation (2) and a value of 306 is obtained. Using a current of 120 μA, a beam energy loss, ΔE, of 0.6 MeV and assuming the beam uniformly distributed in an annular ring between 0.3 and 0.8 cm radius, the value of $T_f - T_g$ is found to be 620°K from equation (4). Using a temperature of 78°K for the gas at the foil gives 425°C for the foil temperature.

From the defining equation given above for the Nusselt number, the heat transfer coefficient is found to be 0.067 watts/cm^2°K. From this it is readily seen that a large temperature difference between the foil and the bulk of the gas is necessary to accommodate the large heat flux. The effective thickness of the laminar sublayer can be calculated from the expression D/(Nu). It is found to be 6.2×10^{-3} cm and it is only because of this small value that the thermal gradient of 1.00×10^5°K/cm can be developed with a temperature difference of only a few hundred degrees to produce the necessary heat flow through the poorly conducting gas.

There remains the justification of the use of the duct equation for the evaluation of the Nusselt Number and the question of the appropriate temperatures to use in evaluating the physical properties of the gas. The Nusselt Number depends, in the case of a gas where the Prandtl Number is nearly unity, primarily on the Reynold's Number which determines the level of turbulence at the foil. Turbulence requires about 10 to 15 duct diameters to develop in a circular duct and about the same distance to die out. Referring to the diagram of the target vessel in Figure 1, it is seen that the deuterium gas enters from the left in a 9.5 mm dia. duct of at least 35 cm length (40 diameters). Thus turbulence is fully established at the level of 9.5 mm dia. duct. This duct empties into a 19.05 mm dia. duct about 7.5 cm (8 diameters of the 9.5 mm duct) from the target vessel. This is approximately the distance for the turbulence to adjust to that characteristic of the 19.05 mm dia. duct. After entering the target vessel, the gas flows in a short annular space of 38.1 mm major and 19.05 mm minor diameter. Since the characteristic diameter of an annular space is the difference of major and minor diameters the annular space has the same effective diameter as the 19.05 mm duct. The gas then blows immediately onto the foil. Therefore the turbulence at the foil has no chance to vary from that for the 19.05 mm dia. duct and thus the value at the foil is characteristic of a long 19.05 mm diameter duct. It is interesting to note that Liljenzin[4] also proposes to treat the foil as a section of the wall of a circular duct.

The temperature to be used in evaluating the Prandtl number is not hard to decide. While good values of the thermal properties of deuterium have not been found for the entire temperature range, an inspection of the variation of the Prandtl number for both hydrogen and helium shows a total variation of only 4.5%. Since the Prandtl number enters the computation to the 1/3 power, the variation with temperature is reduced to only ± 1.5% and may be neglected.

The temperature to be used in evaluating the Reynold's number to determine the level of turbulence is the bulk temperature of the gas and 78°K may be used since the total rise of temperature in the target vessel following complete absorption of the 8.3 MeV beam is only 10°K.

The temperature in the laminar sublayer to be used in evaluating the conductivity of the gas and other factors is another matter. Immediately adjacent to the foil the gas will be at the full foil temperature and the conductivity may be 6 to 8 times the bulk conductivity. As the distance in the laminar sublayer from the foil increases, the temperature and the conductivity will decrease. Since the same quantity of heat must flow thru each differential element of thickness per unit area, the temperature gradient will be smaller near the foil where the conductivity is highest and will increase as the buffer layer is approached. The thickness of the laminar sublayer is, without question, a temporally varying factor much influenced by the turbulent core of the gas. For the case considered here it is estimated that there are frequency components of velocity variation above 10kc/sec. These variations in instantaneous laminar sublayer thickness result in the physical removal of small mass elements of hot gas and thus of heat from the laminar sublayer. Therefore in regions near the buffer layer the flow of heat in the laminar sublayer per unit area due to conduction is reduced and the temperature gradient will now decrease with increasing thickness of the boundary layer. The temperature distribution from the foil will thus have a Gaussian-like shape and the correct mean temperature for evaluating the conductivity factor and thus the foil temperature will be difficult to determine.

On the other hand, as the temperature of the gas rises in the neighborhood of the foil, the viscosity will increase and the density of the gas decrease. Both factors act to dampen and decrease the turbulence in the buffer and laminar sublayer zones and thus act to increase the temporal mean thickness of the latter. This will counteract the effect of increased temperature on the conductivity factor. In fact, the heat transfer coefficient seems to be evaluated surprisingly well by the bulk gas temperature and there is not good agreement among authorities on the need to use some type of mean temperature.

Furthermore the use of any type of mean temperature requires knowledge of the foil temperature to evaluate the physical constants of the gas and make an analytic solution of the problem impossible. The only approach would be an iterative computer solution. It seems preferable to use the bulk gas temperature and the analytic solution and to regard the results as not having high accuracy but rather as giving a qualitative evaluation of the temperature. On the other hand the results do not seem unreasonable and they may certainly be used as a guide to indicate what changes may be useful in improving foil performance.

The above result was obtained for a uniformly distributed beam current density over a limited area of the foil. Assuming the vertical trace of Figure 3 is a linear representation of current density and that the value at the edge of the window (9.5 mm) represented the base or zero current level, a numerical integration indicated the maximum current density was 20% higher than the current density used in the calculations. This would indicate a "hot spot" temperature difference of 745°K and a maximum foil temperature of 550°C instead of the 425°C found for the assumed uniform distribution.

Another factor not previously taken into account is the pulsating nature of the flow of gas from the single acting reciprocating pump. Since the pump operates at about 6 strokes per second, there is a period of about 1/12 second when the flow is produced only by the excess pressure developed in the previous exhaust stroke and the smaller effect of the intake stroke.

Table II. Foil Temperature vs. Flow Factor

Flow Factor	Reynold's Number	Nusselt Number	Foil Temperature, °C
100	1.5×10^5	306	550
80	1.2×10^5	256	700
67	1.0×10^5	221	840
50	0.75×10^5	176	1100
133	2.0×10^5	385	400
100 (D= 0.95cm)	3.0×10^5	533	20
80 (D= 0.95cm)	2.4×10^5	446	60
50 (D= 0.95cm)	1.5×10^5	306	180

The values calculated were for the mean flow of gas which was obtained from the pump displacement and a measured efficiency factor of 75%. The displacement was calculated for the pump with the stroke reduced to 3/4 of the design value and a motor operating at 3480 rpm.

It is difficult to measure the instantaneous flow, but some estimated values can be calculated. Since the foil has very little heat capacity, it can be assumed the foil temperature will be in equilibrium with the cooling for the slow period of the pump. Some representative temperatures for various flow factors are given in Table II where the 20% increase of temperature due to the uneven distribution of beam current of Figure 3 has been included. The values in the table are calculated for the nominal mean flow of 1800 cm^3/sec of deuterium at 78°K and 10 atm. pressure and a beam current of 120 μA.

All the temperatures calculated for gas cooling have assumed that the total energy dissipated in the foil by the beam was carried away by this means. It will now be shown that radiation and conduction make a small or negligible contribution to cooling of the foil. Using Stefan's 4th power law, an emissivity factor for Havar (0.13) equal to that of cobalt and nickel at 500°C and considering radiation from both sides of the foil, the radiation loss at 700°C is only 2.5% of the loss from gas conduction. Thermal conductivity factors have not been available for Havar, but estimating the value from the known electrical conductivity, a loss of 5% of that for convection is obtained for a uniformly distributed beam. Liljenzin[4] gives a factor of 3 less for a single Gaussian beam shape. The double Gaussian observed in Figure 3 would fall between the two extremes and the total loss by other modes would be less than 5% of the convection loss. At 1100°C the radiation loss will more than double, but reasons will be advanced for not regarding that temperature as valid.

The temperatures calculated and summarized in Table II do not seem out of line with expectations when it is considered that the foil was operated close to its maximum beam current limit. The maximum temperature recommended by the producer of Havar is 435°C, but a window of comparable strength was operated at 6 atm. while hot enough for the glow to be seen on a closed circuit T. V.[6]. The temperature would be estimated to be in the range of 600-800°C. The results for a flow factor of as low as 67% are seen to be not out of line with this observation. On the other hand the 1100°C temperature for 50% flow factor is too high (melting is estimated at 1300-1400°C). Also the flow factor is probably too low considering the high flow impedence that is known to exist in the transfer lines.

Indications for Improvement

In a system (the window foil) where the heat capacity is nearly zero, any steps that can be taken to limit temperature excursions will be obviously effective. One such step would be the provision of ballast tanks at the intake and exhaust ports of the reciprocating pump. Another step would be conversion of the pump to a double acting system using the space above the piston and suitable valving. The former would have the disadvantage of a large increase in deuterium gas inventory.

Increase in the size of the transfer lines could increase the flow through increased pump efficiency but too large an increase could remove too much of the smoothing effect of transfer line impedence on flow ripple and allow the minimum flow rate to drop below present values. Some increase would permit the pump stroke to be increased to the full design value. The 133% flow factor shows what could be expected for the mean flow for this condition. Then the value shown for 100% would represent a 75% flow factor for the increased stroke. An improvement of about 150°C is seen for both the mean and minimum flow rates with this change.

The most dramatic improvement would appear to come about through a reduction of the short section of 1.905 cm duct at the entrance to the target vessel (see Figure 1) to 0.95 cm. Then the gas would flow through the smaller sized duct all the way into the target vessel. The increase in impedence of the transfer lines would be negligible. The increased Reynold's number at the foil and the smaller duct diameter have a large effect on foil temperature as shown by the last three entries in Table II. Some loss of turbulence would occur in the annular part of the flow pattern but this is so short that most of the improvement shown in the table should be realized.

Another means that might be effective in reducing foil temperature would be the installation of a turbulence producing screen in the throat of the target vessel immediately adjacent to the foil but out of the way of the beam. Again decreasing foil thickness would decrease energy loss and temperature but at the expense of foil strength. A thickness of 1.0×10^{-3} cm would be adequate to support the cold gas. Whether or not it would take more beam current would depend on the shape of the tensile strength vs temperature curve.

Mechanical Considerations for Circular Havar Membrane

Thin circular windows can be treated by membrane theory when the thickness, h, is small compared to other dimensions, when linear elements which are normal to the undisturbed middle surface

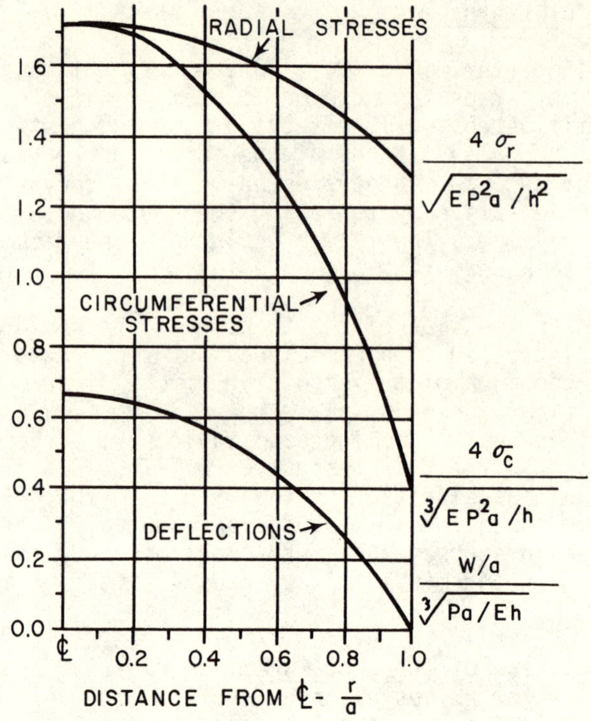

Figure 4. Stresses and deflections of a circular membrane under uniform pressure within the elastic limit.

remain straight and normal to the deflected middle surface of the membrane; and when bending stresses are small and can be neglected. When the stresses are below the elastic limit and the load is a uniform pressure, the radial and circumferential stresses and the deflections are given by the curves in Figure 4. The symbols used there have the following meaning:

- r = Radius to any point on the membrane - cm
- a = Radius to edge of membrane - cm
- P = Pressure - Pascals
- E = Young's modulus - Pascals
- h = Thickness of the membrane - cm
- σ = Radial stress - Pascals
- σ^r = Circumferential stress - Pascals
- W^c = Deflection of the membrane - cm

From the figure it is seen that the radial and circumferential stresses are both at the maximum value and are, in fact, equal at the center of the window. The stresses then are given by:

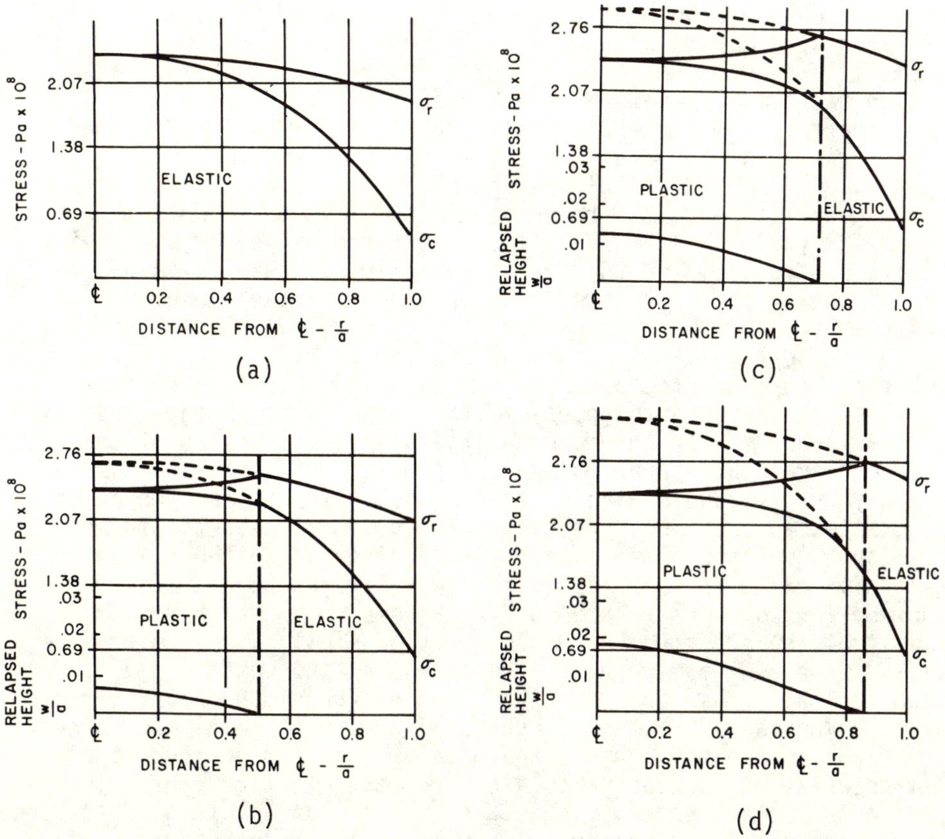

Figure 5. Distribution of stresses in a steel membrane under various degrees of pressure loading:

a) Pressure at the elastic limit; $\sigma_r = \sigma_c = 2.4 \times 10^8$ Pa at r/a = 0.
b) Pressure increased approximately 19%.
c) Pressure increased approximately 35% above (a).
d) Pressure increased approximately 55% above (a).

$$\sigma_c = \sigma_r = 0.43\, E^{1/3} \left(\frac{pa}{h}\right)^{2/3} \qquad (7)$$

Using this relation to calculate the maximum stress at the room temperature bursting pressure of 15 atm. (15.2×10^5 Pa) quoted earlier, an ultimate stress of 2.77×10^9 Pascals is obtained. For the 21.8 atm. bursting pressure observed at 77°K, the ultimate stress at the center is calculated to be 3.56 x 10^9 Pa. No figures are available from the manufacturers for the tensile strength at liquid nitrogen temperature, but the value given at room temperature is 2.2 x 10^9 to 2.4 x 10^9 Pa. This is considerably below the calculated value of 2.77 x 10^9 Pa. The manufacturer also quotes a yield strength of 2.1 x 10^9 to 2.3 x 10^9 Pa. If the foil reaches the plastic range before coming to its ultimate strength, it can plastically deform to a more favorable shape which will support the higher pressure without the stress rising to the bursting limit. This is illustrated in an analogus way by some work published on steel membranes(9). Figure 5a shows the stress curves for a steel membrane when it is stressed to the elastic limit of 2.35 x 10^8 Pa. Figure 5b shows what happens when the pressure is increased about 19% so that part of the membrane becomes plastic. The plastic range extends out to about 50% of the membrane radius. The stresses at the center remain at approximately the previous values, but rise slowly to the edge of the plastic region. The dashed curves show the stresses that would be calculated using the elastic formula and show that these are substantially higher than those that actually exist in the plastic region. Figures 5c and d show curves for approximate pressure increases of 35 and 55%. The limit of the plastically deformed region slowly moves toward the fixed edge of the membrane. The stress at the center remains substantially the same, but that at the edge of the plastic region slowly increases. It would eventually reach the ultimate strength of the membrane and rupture would occur. It can be noted that the stresses at the center calculated from the elastic theory rise far above those that actually exist. It is probably not valid to draw more than quantitative conclusions in applying these curves to the case of Havar. It does seem plausible, though, that plastic deformation can explain the foil's ability to support pressures above those calculated from the ultimate tensile strength. The foil thus takes a more favorable shape as the pressure rises until eventually the ultimate yield stress is reached.

Conclusions

Havar is a remarkably strong material and, in the form of thin foils, provides an excellent material for cyclotron target windows. Some of the ways of handling this material under high pressure and high beam currents have been discussed. Means for increasing the allowable beam current by more than a factor of 3 have been demonstrated. Through an analysis of the cooling of the foil by the cold deuterium gas, ways have been suggested to further increase the beam current.

Additional experimental observations that might be useful come to mind. The foil might be varied in thickness by ± 20%

from the value of 1.27×10^{-3} cm used so far. It would be valuable to have bursting experiments run at elevated temperatures with the window simulating jig. Means of observing the window under beam bombardment and ways of evaluating its temperature would be useful.

Other foil materials have not been investigated. Not many show any promise of surpassing Havar. One material that does look interesting is a titanium alloy[10]. It has a much lower density so that an 80% increase in foil thickness could be used without an increase in beam energy loss. The lower value of the Young's Modulus is also helpful, allowing the window to deflect into a more favorable shape. At the Havar bursting pressure of 15 atm., the calculated maximum radial stress in the titanium alloy would be 1.53×10^9 Pa. This exceeds the quoted tensile strength (as a minimum) by an amount only slightly larger than that for Havar. Much would depend on the shape of the curve of tensile strength vs temperature. Titanium alloys are known to have good properties in this respect.

REFERENCES

(1) Kuchnir, F. T., Waterman, F. M., Skaggs, L. S., Vander Arend, Vander Arend, P. C., and Stoy, S., Design of a Cryogenic Deuterium Gas Target for Neutron Therapy, IEEE 76 CH 1175 9 NPS, 513, 1977.
(2) Kuchnir, F. T., Waterman, F. M., and Skaggs, L. S., A Cryogenic Deuterium Gas Target for Production of a Neutron Therapy Beam with a Small Cyclotron, Proceedings Third Symposium on Neutron Dosimetry in Biology and Medicine, EUR 5848 DE/EN/FR, Brussels - Luxemburg, 1978, p. 369.
(3) Kuchnir, F. T., Skaggs, L. S., Elwyn, A. J., Mooring, F. P., and Frigerio, N. A., Design of a Neutron Therapy Facility for a 30-inch Cyclotron, AIP Conference Proceedings 9, 638, 1972.
(4) Liljenzin, J. O., The Temperature of Thin Foils in Ion Beams, Lawrence Berkeley Laboratory Report, LBL-1912, June 1973.
(5) Benaroya, R. and Ramler, W. J., Deflection Coil for an External Accelerator Beam, Nucl. Instr. and Meth. 10 (1961) 113.
(6) Private communication, Dale K. Wells, King Faisal Specialist Hospital and Research Centre, Riyadh, Saudi Arabia; formerly with Medi-Physics, Inc., Emeryville, California.
(7) Knudsen, J. G. and Katz, D., Fluid Dynamics and Heat Transfer, McGraw-Hill, New York (1958).
(8) McAdams, W. H., Heat Transmission, McGraw-Hill Kogakuska, Ltd., Tokyo (1954).

(9) Stevens, H. H., Jr., Behavior of Circular Membrane Stretched Above the Elastic Limit by Air Pressure, Proceedings of the Society for Experimental Stress Analysis, Vol. ii, Addison Wesley 1944, p. 139.
(10) Titanium Alloy Ti - 4Al - 3Mo - IV. Density = 4.5 gms/cm^3; modulus of elasticity - 1.14 x 10^{11} Pa; tensile strength (aged) = 1.24 x 10^9 Pa (minimum); yield strength (aged) = 1.07 x 10^9 Pa (min.). Data from Hamilton Technology.

HUMAN FACTORS OF SAFE TARGET HANDLING

L. R. Smith, J. Cameron, and M. Nappi

New England Nuclear Corporation
601 Treble Cove Road
N. Billerica, MA 01862

ABSTRACT

The purpose of this paper is to discuss the higher dose commitment of cyclotron target handlers as compared to other radiation workers. Target handling technical details have not been described since they would detract from the main concern, which is the human factor of personnel exposure. Instead, characteristic reasons for high personnel exposure are presented along with suggested methods for reducing these exposures. Dose commitment trends at NENC are presented to show the impact of implementing suggested improvements. This discussion is intended to be of general interest and is illustrated by examples of operating experiences using three thirty-inch commercial isotope production cyclotrons.

INTRODUCTION

There are six contributing factors for commercial cyclotron workers' poor radiation exposure history. These may be divided into two categories vis.:

1. a. Commercial cyclotrons are said to be difficult to operate without incurring personnel exposure.
   b. Cyclotron workers are often unconcerned about their exposure.
   c. Regulatory and health physics controls are insufficient.

2. a. Preparation of the target.

b. Placing targets on, and removing targets from, the cyclotron.
c. Processing the irradiated target.

Preparation of the Target

Target plating operations require careful, intimate handling to ensure that target products are not removed during the irradiation or transfer phases. The scarcity and expense of certain enriched materials suitable for radioisotope production may necessitate recycling. The technologists handling the target are likely to give less credence to the smaller quantities of activity remaining on a target after process and decay. They further tend to disregard the basic physics of inverse square attenuation and subject themselves to unnecessary and potentially excessive extremity exposure. To accurately determine these exposures, health physicists should periodically audit the operation while the technologist wears an array of finger dosimeters (e.g. thermoluminescent dosimeters). The technologist should be informed of the results of the audit so that he will recognize the potential for high local exposure. Once aware of hazards, the technologist can take steps to prevent dangerous and unnecessary exposure by engineering appropriate controls: applying time, distance and shielding concepts to the operation.

Our experience has shown that the most effective method of bringing the technologist to this level of consciousness is by having a skilled health physicist supply him with direct measurement data along with Polaroid action photographs. We further feel that this technique should be employed in other operations to give the technologist the necessary information to guide him toward alternative handling methods.

Target Installation and Removal

High dose commitments are incurred during target installation and removal. This situation is both more serious than target preparation exposure and more difficult to resolve. The target changer is besieged on three fronts. He is required to place a moderately radioactive target in a machine which has high residual radiation fields, and in a room which may contain other sources of radiation. Unless a remote removal mechanism is employed, the technologist must also remove the irradiated target for further processing.

Flow-through targets and remote automatic methods for handling intensely radioactive materials have long been used in the nuclear industry. However, there are several reasons why target handlers do not use these methods but instead remove the target manually. These reasons include:

1. Remote systems are expensive.
2. Many cyclotron vaults were not designed to accommodate a suitable system.
3. Remote systems impede access to the cyclotron for maintenance purposes.
4. Devices installed in a cyclotron vault may be subject to radiation damage.
5. Remote systems require maintenance.
6. Production must be interrupted to install and maintain remote systems.
7. Remote systems require a redundant backup.
8. Targets are not always designed to be easily handled in remote systems.

It is interesting to note that a lot of attention is applied to designing targets which can be accurately located in an accelerator beam, which do not lose activity and are efficiently cooled during irradiation. Much less effort is directed into designing targets for easy handling. It is obvious that a target should be custom-designed to allow remote manipulation, but what is often forgotten is that proper design is also essential to facilitate manual manipulation. Even when using remote systems it may only be practical to recover a target manually when the remote system fails. The target should therefore be designed to facilitate handling in both situations.

NEN decided long ago to invest in a remote target changing system, and our engineers are currently working on its installation. For many years a manual method was used. Distance was employed by means of a ten-foot handling pole. A shield wall protects the major part of the body during the operation. Although the all-male team were more concerned about gonad exposure, studies have consistently shown the eyes to be the critical organs for all target operations in the vault. This finding is characteristic of NEN's operations, but it probably applies to similar facilities. Responsible supervisors and health physicists should monitor this possibility periodically and establish eye-whole body exposure ratios for all routine operations.

Another interesting facet of these target-handling studies is that the presence of a health physicist monitoring each operation causes personnel to increase their efficiency, resulting in less time taken and lower dose commitments. Dose commitments would be even lower if unplanned mishaps could be eliminated. These include target jamming, target box windows that fail to open, dropped targets, tools not on hand, and failure to follow standard operating procedures. Mishaps such as these plague as many as thirty percent of the operations at NEN and significantly increase the average dose commitment.

Supervisors must be encouraged to plan for the unexpected. Although difficulties can arise in complex environments or processes, proper planning could reduce their frequency. Planning for the unexpected should be included in training. New technologists should practice proper procedures on simulated models until they can perform with speed and dexterity. Proper planning includes eliminating the unexpected by documenting, posting and learning carefully tested procedures, and ensuring a clear, uncluttered work environment.

One should not leave target installation and removal considerations without mentioning cyclotron maintenance operations. Many NEN target handlers are also employed to assist in maintenance operations, an important source of exposure. Preplanning, training, supervision and health physics auditing are essential ingredients to minimize dose commitments. Another important control, becoming less palatable in anti-authoritarian societies, is to ensure that technologists exercise strict discipline when close to the cyclotron. This means they must respond promptly to orders, concentrate their efforts, and leave high exposure rate zones whenever they are not working at top speed.

The employment of appropriate shielding is a frequently underrated aid. Supervisors often believe that it is impossible to shield a cyclotron since the whole machine is activated and contaminated. There is widespread belief that personnel incur a dose commitment when placing and removing shields which cancels any dose saved by the shield's presence. Although these arguments may be applicable for many machines, they certainly do not apply to those at NEN. We have used many different shields to some effect. Recent experiments have shown that the high exposure rates encountered are mostly due to direct radiation from intensely activated hill piece leading edges and the central region. Assumed "widespread activation" was found to be due to secondary radiation Compton-scattered from objects exposed to these intense local sources. A small amount of shielding judiciously placed in seconds can substantially reduce exposure rates while leaving the rest of the machine open to work on. Many maintenance operations do not require an engineer to lean far into the machine, and it would be quite feasible to reduce exposure by approaching the machine on a shielded cart.

To ensure these suggestions have maximum impact, the work area must be brightly illuminated and the operator should have a place at hand to store his tools.

The condition of the machine directly affects the dose commitments of target handlers and maintenance mechanics. It is noticeable that in one of our cyclotrons unusual beam loss is causing harmonic coils to become intensely activated. This and similar faults inevitably cause unnecessary exposure and should be corrected when possible.

Personnel exposure is also affected by the amount of down time that can be allocated to allow the residual activity to decay before maintenance work is performed. Although seldom achieved in practice, supervisors should try to schedule long down times to perform preventive maintenance rather than to stop the machine frequently for short periods just to attend to fault conditions.

The literature contains many papers describing contamination control in cyclotron vaults. At NEN, although contamination control is difficult, it represents a nuisance rather than a hazard. Due to regularly scheduled surveillance and prompt action to evaluate and contain breaches in contamination control, internal contamination is hardly detectable and makes a negligible contribution to the total dose detriment. It is good health physics practive to ensure that internal contamination is insignificant.

Target Post Irradiation Processing

Targets are chemically processed in nine-inch thick lead cells provided with eighteen-inch thick lead glass windows and remote over-the-top manipulators. Defective cell designs have caused unnecessary personnel exposure in the past at NEN. Exposure was mostly due to manual cell loading and waste removal, accidental releases of processed activity to unshielded HEPA filters located in the same room, and photon streaming through poorly engineered penetrations. We have learned from these mistakes and are replacing these old cells with a new generation allowing for remote loading and improved air handling capabilities.

Legislative Controls

It is no secret that one reason why personnel dose commitment might be high is because they are allowed to be. In many countries legislators have adopted the maximum permissible annual occupational exposures recommended by the I.C.R.P.[1] Occupationally exposed personnel are permitted to contract up to 3 Rems per quarter provided that the average dose equivalent commitment does not exceed 5 Rems per year after eighteen years of age. The

3 Rems per quarter relaxation is intended only for occasional use when every effort has been made to keep exposure to as low as reasonably achievable (ALARA). In practice, however, many occupationally exposed personnel improperly interpret this relaxation as an open invitation to relax their exposure controls.

In the U. S. A., there are two phenomena which may change this situation in the near future. The U. S. Nuclear Regulatory Commission has offered for comment a proposed rule change[2] which intends to disband the relaxation. Although personnel shall continue to be allowed up to 3 Rems whole body exposure in a given quarter, they shall be limited to 5 Rems each and every year. It is interesting to note that the N. R. C. did not disband the 3 Rem per quarter limit, since they deemed that to do so would only lead to unnecessary dose sharing by people who lack the vision to employ more appropriate controls.

Dose sharing should be avoided, since when an operation is shared among several persons often the total dose commitment will be higher than if only the most skilled worker performed the operation alone. Personnel should be selected to perform high exposure rate operations according to their skill and efficiency only, other measures being taken to ensure that their dose commitments remain ALARA and do not exceed legal limits. In order to ensure that these proper controls are implemented, the responsible health physicist should analyze the total man-rems incurred by each operation. The strategy employed to reduce these total man-rems should be assertively pursued, planning being coordinated with the managers responsible for the total operation.

The second change that is already upon us is the increased public concern about risk from low-level radiation. This has partially resulted in new employees' being much more concerned about their exposure than before. They are demanding tighter controls, and this general change in attitude can be expected to lead to dose reduction in the future.

NEN: Radiation Control Experience

For proprietary reasons it has not been possible to give detailed accounts of safety techniques, since they are an intimate part of the manufacturing process. The issues expressed here are a mixture of tested experiences, changes currently in the design and installation stage, and even a few future considerations. At the beginning of 1979 we discussed a very interesting proposal to build a special training laboratory, including a mock cyclotron for testing remote devices and shielding design, and to train technologists and health physicists. Whether this proposal can be

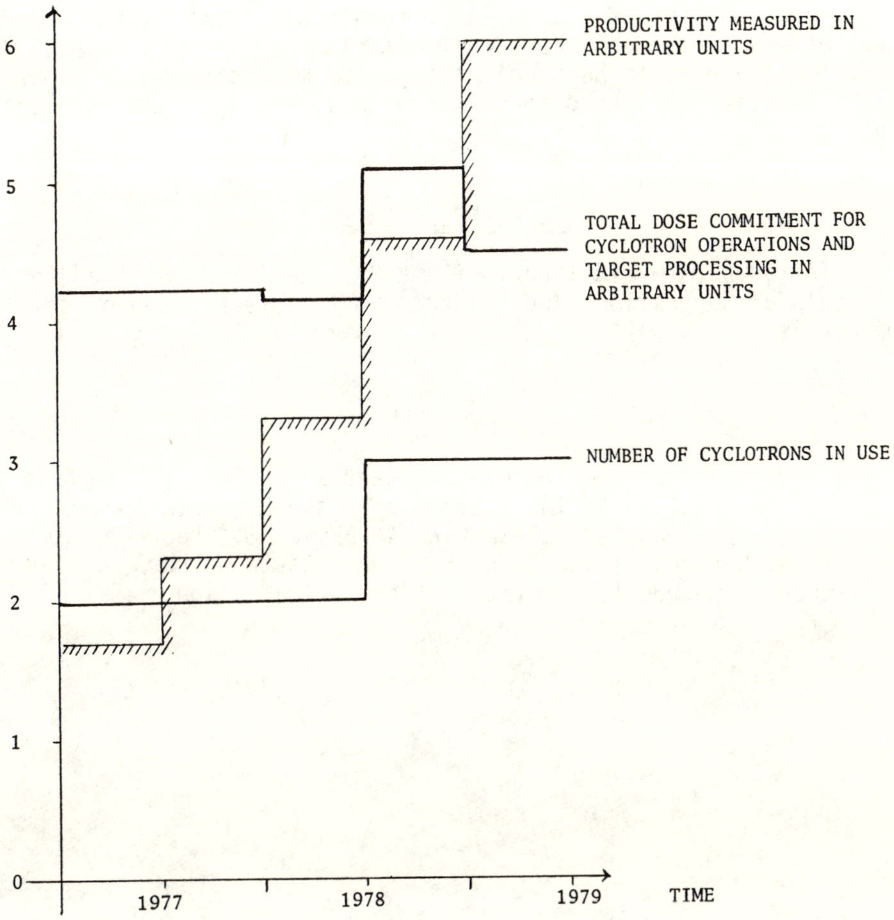

Figure 1. Comparison of trends in:

a. Total personnel dose commitments incurred during cyclotron and target processing operations.
b. Productivity.
c. Number of cyclotrons in use.

made cost-effective depends greatly upon the progress of other improvements.

The general improvement in NEN Cyclotron Group dose commitments during the past two-and-a-half years is illustrated in the preceding figure. Note that the total man-rems for all cyclotron and target processing operations have remained relatively stable in spite of substantial increase in productivity during the past two-and-a-half years. This is encouraging and is due to the

Cyclotron Group's efforts to limit total dose commitment. Progress had not been slowed by the law of diminishing returns; indeed, dose reduction appears to be accelerating. By making exposure control a high priority, we intend to continue this progress long into the future.

ACKNOWLEDGEMENTS

The authors would like to thank Mr. Charles B. Killian for his helpful suggestions and guidance in the preparation of this manuscript.

REFERENCES

(1) I. C. R. P. 2. Report of Committee II on Permissible Dose for Internal Radiation, Pergamon Press, London, 1959.
(2) 10 CFR 20 Proposed Rule Making 44 FR 10388. February 20, 1979, Notices, Instructions and Reports to Workers: Inspection Standards for Protection Against Radiation.

METHODS TO REDUCE CONTAMINATION IN TARGETS PREPARED BY VACUUM DEPOSITION

G. E. Thomas, S. K. Lam, and R. W. Nielsen

Argonne National Laboratory
Argonne, Illinois 60439

For some time, both the experimenters and we target makers have been concerned about the source of impurities which were found from experimental data obtained using targets prepared by vacuum vapor deposition. These impurities may arise from the process of producing the target, the separated isotope used in the evaporation, or contaminants introduced during the experiment. Some impurities such as carbon, oxygen and nitrogen are likely to be introduced into the target during evaporation, as a result of residual gases and contaminants, such as diffusion pump oil, present in the vacuum system. The present study was initiated to determine the source of impurities found in our targets. We would like to mention some of the experimenters who were kind enough to cooperate with us by bringing to our attention these impurities whenever they found them and, in addition, letting us use some of their data: D. F. Geesaman, W. Henning, M. S. Kaminsky, D. G. Kovar, W. Kutschera, G. C. Morrison, M. Paul, and S. J. Sanders.

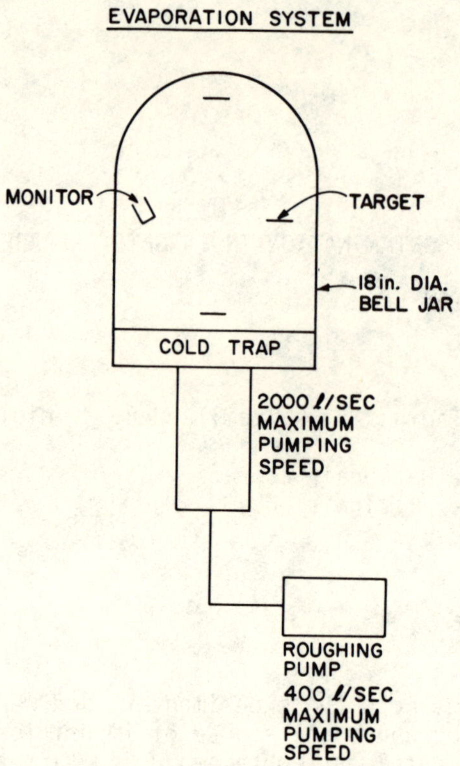

Figure 1.

Our evaporation system, a Varian NRC[1], is shown schematically in Figure 1. It uses an 18-inch glass bell jar, a NRC VHS6 diffusion pump having a 2000-liter per second maximum pumping speed, a 400-liter per minute Welch 1397B roughing pump and a large-capacity liquid nitrogen cold trap. The liquid nitrogen cold trap improves the vacuum by about a factor of ten as compared with a refrigeration-type trap. A Kronos crystal thickness monitor[2] was used in all evaporations.

From Deutschmann[3], we find that the number of atoms striking the surface per second is given by the following equation:

$$\gamma = \tfrac{1}{4}\, n\, V_a,$$

$$\gamma = 3.513 \cdot 10^{22} \frac{P}{(MT)^{\frac{1}{2}}} \ \text{cm}^{-2}\ \text{sec}^{-1},$$

where M = mass (AMU), T = degrees Kelvin, and P = pressure in torr. For a typical evaporation system at 10^{-6} torr, the number of oxygen atoms striking a given substrate would be 1 x 10^{14} per square cm per second. A monolayer of oxygen consists of ∿ 8 x 10^{14} molecules per square cm per second. In other words, there is a monolayer of O_2 striking the substrate about every 2 seconds. At a vacuum of 10^{-6} torr, 0.02 micrograms per square cm of oxygen would potentially be deposited on the surface of the substrate per second, if the oxygen sticking probability was unity. It should be pointed out that this amount of oxygen will not all be deposited on the substrate, but the above estimate does indicate that there is a possibility of considerable oxidation of the target prepared by vapor deposition in a vacuum of 10^{-6} torr, which is usually considered to be suitable for an evaporation. When an evaporation is made in a vacuum system evacuated with an oil diffusion pump, dissociation of hydrocarbons (from the oil vapor) may increase the carbon contamination problem.

Some examples of experimental data obtained with the targets which we have prepared, will be presented. This will give some idea of both the kind and magnitude of the impurities wich have been found. Boron-11 targets were prepared for Morrison et al(4) for their experiments in elastic scattering of ^6Li from ^{11}B at the

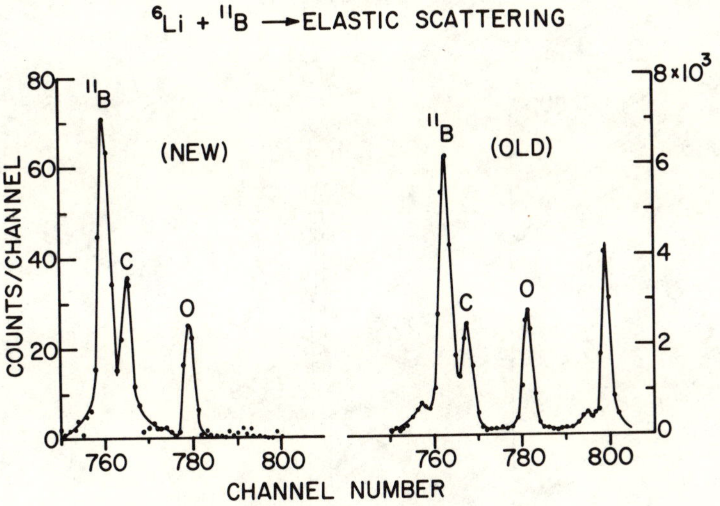

RATIO	OLD	NEW
C/B	0.40	0.51
O_2/B	0.45	0.36

Figure 2.

Argonne FN tandem accelerator. It was found that the first set of targets, made some time ago, contained many impurities, as can be seen in the data shown in Figure 2. In addition to carbon and oxygen, other impurities were found. A second set of targets were then evaporated in a vacuum of 10^{-6} torr with a high purity sample of ^{11}B, purchased from Eagle Picher[5]. The data obtained with this new set of targets, shown in Figure 2, clearly indicate that the impurities other than carbon and oxygen disappear. This observation is attributed to the improved purity of the ^{11}B sample used in the evaporation. The oxygen and carbon contaminants are, in part, due to the residual gases and oil vapor present in the vacuum chamber during the evaporation. For the new target, the ratio of the intensity of oxygen line to that of the boron line is somewhat reduced while the ratio of the carbon intensity to boron intensity is larger. These results suggest that the new target was produced in both a better and cleaner vacuum.

Data taken by S. J. Sanders et al[6] is shown in

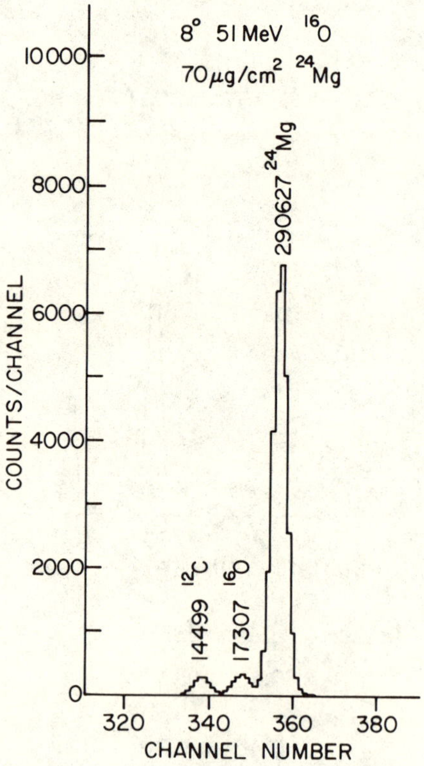

Figure 3.

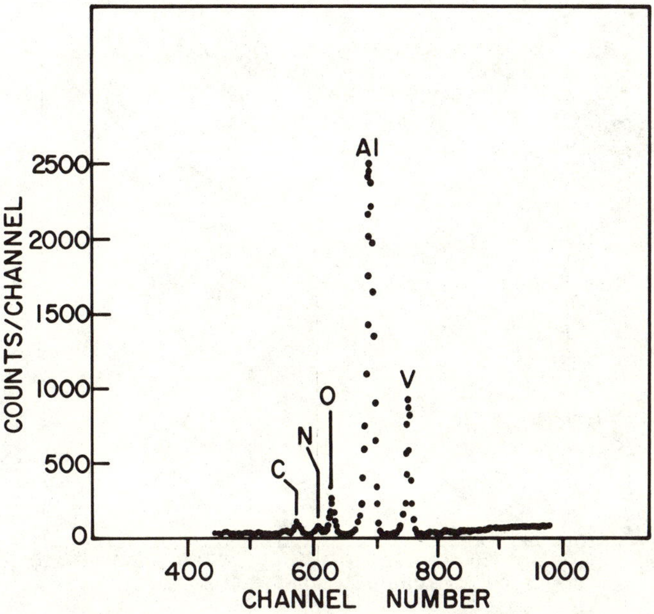

Figure 4. 2-MeV Van de Graaff, 1.25 MeV Protons 50 monolayer Vanadium on 50 $\mu g/cm^2$ Al Electron Gun Evaporation.

Figure 3. Again, this experiment was performed at the tandem accelerator using 51-MeV oxygen-16 particles with the Compton spectrometer. The targets were 70-microgram per cm^2 magnesium-24 foils produced jointly by our facility and that of F. Karasek's[(7)]. These were the targets which first started us thinking seriously about the source of oxygen and other impurities. The oxygen impurity found in these targets made it difficult for the experimenters to analyze their data. Occasionally foils would be produced which had less oxygen impurity than others, but we were not sure of the cause. It was not certain if the decrease in oxygen impurity was a result of the rolling or some other phenomena as rolled targets were not always the lowest in oxygen content.

Recently, some vanadium targets were produced for Kaminsky's[(8)] group for their experiments at the 2-MeV Van de Graaf at Argonne. These targets were prepared using an electron gun to evaporate 50 monolayers thickness vanadium on a 50-$\mu g/cm^2$ aluminum backing. Figure 4 shows some of the Rutherford backscattering spectra of the targets, obtained with their backscattering spectrometer. In addition to the vanadium and aluminum peaks which are characteristics of the target and substrate, peaks of oxygen, nitrogen and carbon are clearly observable. These impurities were undesirable as the oxygen and carbon peaks appeared in the region

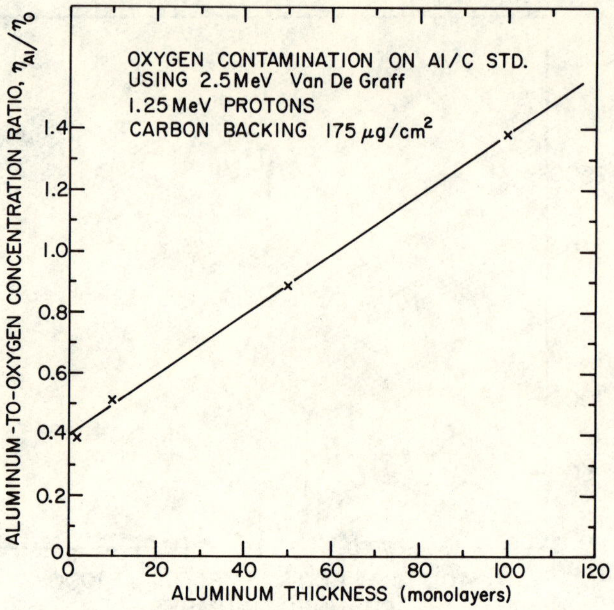

Figure 5. Oxygen contamination on Al/C std. using 2.5 MeV Van de Graaff 1.25 MeV protons carbon backing 175 $\mu g/cm^2$.

of their experimental interest. Furthermore, vanadium targets having the highest possible purity were a necessity because these targets were to be used as standards to calibrate other targets obtained by means other than vacuum vapor deposition. It is believed that the oxygen, nitrogen and carbon impurities were caused by the presence of residual gases in the vacuum chambers during the evaporation of the aluminum substrate and the vanadium target.

Data obtained with the aluminum targets on carbon backing that we prepared for Kaminsky's group(9) further illustrate the problem of oxygen contamination due to residual gases present in the vacuum system. Three aluminum targets were made in a single evaporation by placing three carbon substrates at distances of 7.0, 22.1, and 70.0 cm, respectively, from the source of evaporation. Figure 5 shows the aluminum-to-oxygen concentration ratio vs. the thickness of the aluminum target. It is seen that this ratio is directly proportional to the aluminum thickness. This result can be explained by the difference in the aluminum arrival rate and a constant arrival rate of oxygen at the substrates. It should be noted that a constant arrival rate of oxygen is merely an indicator of the residual oxygen partial pressure in the evaporator vacuum. It is unlikely, based on Deutschmann's(3) effusion equation, that

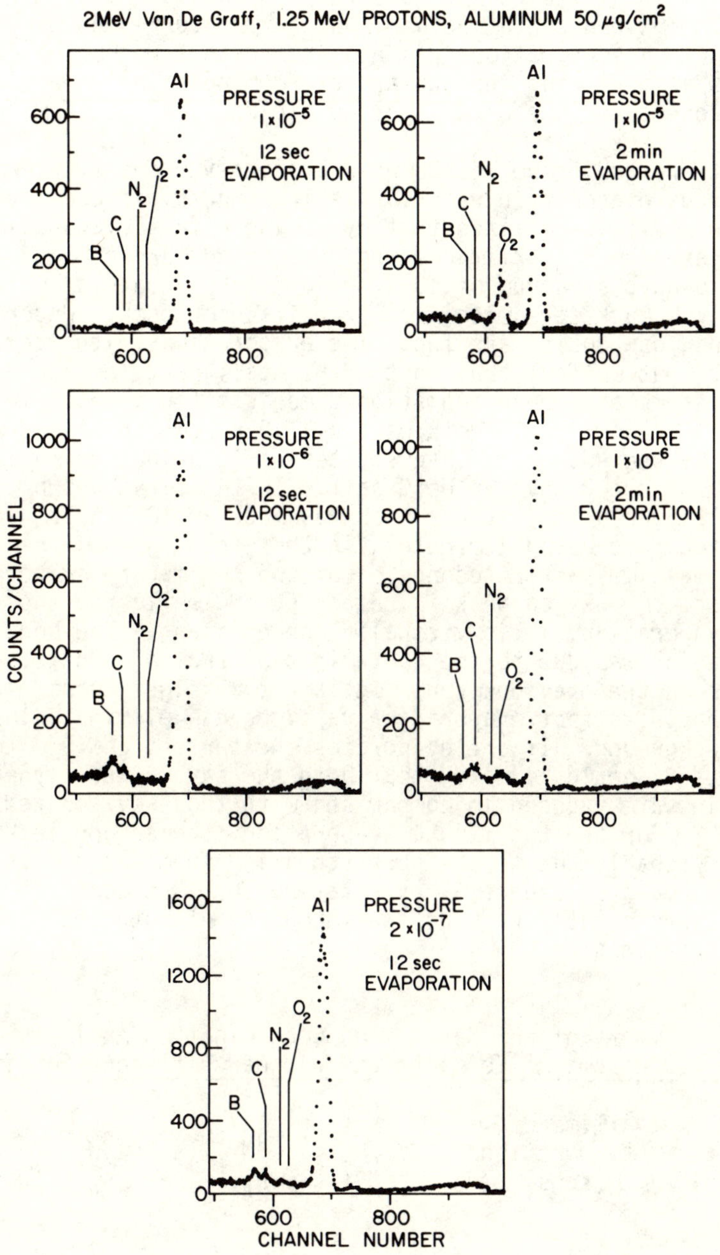

Figure 6. 2-MeV Van de Graaff, 1.25 MeV protons, aluminum 50 μg/cm².

the evaporated aluminum would pick up significant amounts of oxygen as it traveled to the substrate in the vacuum used ($\sim 10^{-6}$ torr). The oxygen contamination must have occurred on the target surface during evaporation. As aluminum hits the substrate, oxygen is also hitting the substrate, forming an oxide as it is being deposited or the oxygen may be occluded in the aluminum.

Further studies were made by preparing various aluminum foils at different evaporation rates as well as under different vacuum conditions. It is realized that by having a fast evaporation, one may decrease the uniformity of the evaporated surface. It would be nice to have a vacuum of 10^{-8} or 10^{-9} torr in a system in which there was no back streaming from the diffusion pump. Under these conditions, one could make long-time evaporations with many less impurities but we feel that most of us still have less than ideal vacuum systems and the evaporation time is often a determining factor. The aluminum targets were all 50 μg/cm² thick and were analyzed by the Rutherford Backscattering technique using 1.25 MeV protons from the 2 MeV Van de Graaff. In the first group of data shown in Figure 6, the evaporation was made at 10^{-5} torr, with the evaporation time being 2 minutes; in the second group, the evaporation time was 12 seconds. For the two minute evaporation, one can see the oxygen peak is rather strong while for the 12 second evaporation, it is virtually non-existent. The boron contamination was due to the contamination from a boron evaporation made in the previous run. It is rather significant, we feel, that by having a fast evaporation rate one can eliminate the oxygen background. If one evaporates the targets at 10^{-6} torr pressure, it should be noted that both the carbon and oxygen peak intensities are reduced in comparison with those evaporated at 10^{-5} torr. For the two minute evaporation, the carbon peak is enhanced probably due to the back streaming from the diffusion pump. As could be expected, evaporating for 12 seconds at 2×10^{-7} torr greatly reduces the intensity of both the carbon and oxygen contamination.

Table 1. Summary of RBS Results on Contamination Analysis for 50 μg/cm² Al Foils Prepared Under Different Conditions.

P (torr)	Time of Evaporation (secs)	$\frac{n_C}{n_{Al}}$ (%)	$\frac{n_N}{n_{Al}}$ (%)	$\frac{n_O}{n_{Al}}$ (%)
1×10^{-5}	12	3	3	4
1×10^{-5}	120	3	12	30
1×10^{-6}	12	6	2	2
1×10^{-6}	120	7	3	3
2×10^{-7}	12	7	2	2

Table 1 points out some significant differences of various evaporation times and pressures. For a constant evaporation time, the carbon-to-aluminum ratio gets larger as one has a better vacuum. Likewise, under improved vacuum and for short evaporation time, the oxygen contaminant as well as the carbon contaminant becomes much smaller. By having a better vacuum, one increases the carbon contaminant but reduces the oxygen.

The conclusions of the present work are the following. In particular, for targets such as magnesium, aluminum, calcium and other readily oxidized materials, the following points are important. One needs a clean system, the best possible vacuum attainable, and as short an evaporation time as the desired uniformity of a target will permit. One must decide what is important for a particular experiment. A word about the mentioned e-gun type evaporation: normally it takes a relatively long time for this type of evaporation. For such a long evaporation time, a good vacuum is even more of a necessity. The oxygen contamination will become more severe as fewer atoms of the evaporated material strike the substrate per unit time. A very good vacuum will certainly reduce the oxygen contamination. Relatively little developmental work has been done using the cryopump system, but it has potential for reducing many contaminants, particularly carbon. There are several problems associated with these systems, but they are not insurmountable. Again, higher purity targets are going to become increasingly more important to the experimenter. More time must be devoted to obtaining high purity materials, the type of evaporation produced and the conditions under which we do evaporation condensation target preparation.

REFERENCES

(1) Varian/NRC 3117 vacuum coating system manufactured by Varian, Palo Alto Vacuum Division, 611 Hansen Way, Palo Alto, CA 94303.

(2) Kronos thickness monitor, Model QM321. Distributed by Veeco Instruments Inc., Terminal Drive, Plainview, NY 11803.

(3) S. Deutschmann, Scientific Foundations of Vacuum Technique, second edition, edited by J. M. Lafferty (John Wiley and Sons, Inc., NY, 1962), p. 14.

(4) G. C. Morrison, D. F. Geesaman, W. Henning, and D. G. Kovar, Private communication.

(5) Eagle Picher Industries, Inc., Electronics Division, Miami, OK 74354.

(6) S. J. Sanders, M. Paul, J. Cseh, D. F. Geesaman, W. Henning, D. G. Kovar, C. Olmer, and J. P. Schiffer, Resonant Behavior of the $^{24}Mg(^{16}O, ^{12}C)^{28}Si$ Reaction, to be published.

(7) F. J. Karasek, Rolling of Evaporated Magnesium Isotopes, published in these proceedings.

(8) M. S. Kaminsky and S. K. Lam, Private communication.
(9) S. K. Lam and M. S. Kaminsky, Private communication.

INDEX